T0136329

Multidimensional Modulations in Optical Communication Systems

Multidimensional Modulations in Optical Communication Systems

Silvello Betti

Pierluigi Perrone

Giuseppe Giulio Rutigliano

CRC Press
Taylor & Francis Group
Boca Raton London New York

CRC Press is an imprint of the
Taylor & Francis Group, an **informa** business

First Edition published 2021
by CRC Press
6000 Broken Sound Parkway NW, Suite 300, Boca Raton, FL 33487-2742

and by CRC Press
2 Park Square, Milton Park, Abingdon, Oxon, OX14 4RN

CRC Press is an imprint of Taylor & Francis Group, LLC

© 2021 Taylor & Francis Group, LLC

The right of Silvello Betti, Pierluigi Perrone, and Giuseppe Giulio Rutigliano to be identified as authors of this work has been asserted by them in accordance with sections 77 and 78 of the Copyright, Designs and Patents Act 1988.

ISBN: 978-0-367-43333-8 (hbk)
ISBN: 978-1-032-03385-3 (pbk)
ISBN: 978-1-003-00247-5 (ebk)

Typeset in Sabon
by SPi Global, India

To Barbara
To Maria, Giuseppe, and my parents
To Elena, Luca, Paolo, and Marco

Contents

Preface

Modern telecommunications networks transport, almost in real time, impressive amounts of information, and the only physical means capable of guaranteeing these performances is the optical fiber that, in fact, constitutes the backbone of all modern telecommunications networks, including 5G.

The purpose of this book is to retrace the technical–scientific development of fiber optic transmissions from the beginning to the present day, trying to understand what the future holds.

Chapter 1 describes the technological developments, inventions and devices that have allowed, through various transitions, the passage from the first optical fiber transmissions, characterized by a not negligible attenuation and a small – if compared to today's – transmission bandwidth, up to modern telecommunications in optical fiber, characterized by very low attenuation and very high throughput per each fiber. This is not a constant development since, often, the availability of new technological devices, such as new lasers, amplifiers, and optical fibers, has allowed a rapid development of the technology up to the saturation of available resources, pending the next technological leap.

Chapter 2 describes how some physical properties of optical fiber can be exploited to improve data transmission techniques. In particular, effects acting on the polarization and phase of the optical field propagating along the fiber are analyzed and controlled so as to make it possible to design novel optical communication systems. This approach has then suggested the direction for subsequent technological developments.

Chapter 3 offers a look at some innovative multidimensional fiber optic transmission techniques that appear to offer advantages over conventional transmission techniques in terms of transmission capacity, resilience to disturbances and information protection. These techniques, along with others, could be the new perspective we need to start designing and building the next generation of fiber optic telecommunications.

Introduction

Today's society relies – to an unprecedented extent – on broadband communication solutions with applications such as high-speed Internet access, Internet of Things (IoT), mobile voice and data services, multimedia broadcast systems and high-capacity data networking for grid computing and remote storage. The total Internet traffic has grown from 10^3 Tbyte in 1997 to more than 10^9 Tbyte in 2019 [1, prg 1. Introduction].

The information highways that make these services possible are, almost exclusively, optical fibers [1, prg 1. Introduction]. No other known medium can support the massive demands for data rate, reliability and energy efficiency. The race for ever better performance continues and the capacity of a single fiber has been boosted by several orders of magnitude, from a few Gbit/s in 1990 to hundreds of Tbit/s today. In particular, to construct long-haul optical networks, Standard Single-Mode Fiber (SSMF) in conjunction with the Erbium-Doped Fiber Amplifier (EDFA) has become the bedrock for the growth of the global Internet. SSMF designs have not changed substantially for many years; fortunately, until recently, the intrinsic capacity of SSMF has always been far in excess of what has been needed to address traffic demands.

However, laboratory-based SSMF transmission experiments are now edging ever closer to information theory-based capacity limits, estimated at ~100–200 Tbit/s due to inter-channel nonlinear effects. This fact has sparked concerns of a future "capacity crunch" [2], where the ability to deliver data at an acceptable level of cost per bit to the customer is increasingly outpaced by demand.

Over the past decades, network traffic has been growing consistently between 30% and 90% per year. While packet router capacities, rooted in Moore's Law, have been matching the above traffic growth numbers for decades, high-speed optical interface rates have only exhibited a 20% annual growth rate, and the capacities of fiber-optic Wavelength Division Multiplexing (WDM) transmission systems have slowed down from 100% of annual growth in the 1990s to a mere 20% per year [1, prg 5. Spatial multiplexing].

Commercially deployed WDM systems in 2010 supported ~100 wavelength channels at 100 Gbit/s each for ~10 Tbit/s of aggregate per-fiber WDM capacity. With a 40% traffic growth rate, one should expect the need for commercial systems supporting 10 Tbit/s (super)channels with per-fiber capacities of 1 Pbit/s around 2024. (Note that this does not mean that such systems will be fully populated by that time, which was not the case for the systems available in 2010 either, but the commercial need to start installing systems capable of such capacities will likely be there.) Both interface and capacity targets require optical communication technologies to overcome huge engineering and fundamental obstacles [3].

Of the five physical dimensions that can be used for modulation and multiplexing in communication systems based on electromagnetic waves (Figure I.1), optical core networking technologies commercially deployed today already make full use of time, quadrature and polarization, employing complex quadrature modulation formats, polarization multiplexing, digital pulse shaping and coherent detection.

To further scale interface rates and fiber capacities, it has thus become mandatory to employ parallelism in the only two remaining physical dimensions that are left for capacity scaling: frequency and space.

Transmission demand continues to exceed installed system capacity, and higher-capacity WDM systems are required to economically meet this ever-increasing demand for communication services. There are many considerations that influence technology selection for network operators building modern optical networks:

- fiber capacity
- network cost
- network engineering simplicity
- port density
- power consumption
- optical layer restoration

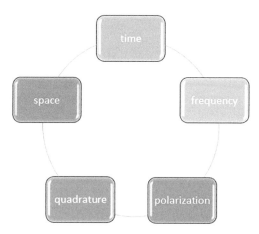

Figure I.1 Five physical dimensions for capacity scaling in communication systems using electromagnetic waves.

Coherent transceivers are very successful, as they have lowered the network cost per bit as the number of transported bits increases. Current state-of-the-art optical coherent transceivers use phase and amplitude modulation and polarization multiplexing on the transmitter side; and coherent detection, Digital Signal Processing (DSP) and high-performance Forward Error Correction (FEC) at the receiver. Coherent detection of amplitude and phase enables multipoint modulations to be applied digitally. The main commercial ones are binary Phase-Shift Keying (PSK), quadrature PSK and 16-Quadrature Amplitude Modulation (QAM), allowing 50 Gbit/s, 100 Gb/s and 200 Gbit/s, respectively [1, prg 6. Coherent transceivers]. The application for each of these, in order, is as follows: submarine links, terrestrial long-haul systems and metro/regional networks.

In the past decade, the development of advanced modulation formats and multiplexing schemes has led to a tremendous increase of the available capacity in single-mode optical fibers. Ten years ago, commercial systems were mostly based on the simple binary on–off keying modulation format with a bit rate of 10 Gbit/s per wavelength channel with a frequency spacing of 50 GHz or 100 GHz between wavelength channels. Today, commercial products offer bit rates of up to 200 Gbit/s per wavelength channel occupying a bandwidth of 37.5 GHz. Such high spectral efficiency is enabled by applying Polarization-Division Multiplexing (PDM), advanced multilevel modulation formats such as M-ary QAM, digital spectral shaping at the transmitter, coherent detection and advanced FEC [4]. In effect, the spectral efficiency of commercial systems has increased by a factor of 27 from 0.2 to 5.3 bit/s/Hz [1, prg 7. Modulation formats].

Achieved line rates per wavelength channel for systems employing digital coherent detection and for Intensity Modulation Direct Detection (IM/DD) systems are summarized in [1, prg 7. Modulation formats].

The current record of 864 Gb/s on a single optical carrier was achieved by employing the PDM-64QAM format at a symbol rate of 72 GBd [5].

The first IM/DD system, which was entirely based on electronic Time-Division Multiplexing (TDM) and operated at a bit rate of 100 Gbit/s, was reported in 2006 [6]. In the following years, research shifted to coherent systems until new solutions for data center communications at 100 Gbit/s and beyond were required. Currently achieved symbol rates and spectral efficiencies are summarized in [1, prg 7. Modulation formats].

For submarine and long-haul systems, maximization of the spectral efficiency for distance product is fundamental, while cost and power consumption are not as critical as in metro and access markets. By supporting several modulation formats, todays flexible transponders are able to adapt bit rate, spectral efficiency within certain limits.

Finally, nonlinear impairments pose a major challenge, which can be partly overcome by intelligent design of modulation formats [7]. The bit rate transmitted per optical carrier needs to be increased while power consumption is limited due to small form factor pluggable modules. Furthermore, the cost per transmitted bit needs to further decrease. Thus, the challenge is to support high bit rates per optical carrier while using low-cost components and low-power electronics. Similar challenges are found in access networks, where the Optical Network Unit (ONU) is extremely cost sensitive and needs to operate at very low power consumption. The desired bit rate per residential user is currently in the order of 100 Mbit/s to 1Gbit/s. However, for next generation heterogeneous access networks supporting optical 5G mobile front haul, the required bit rate per ONU is expected to increase tremendously. Support of such data rates in Passive Optical Networks (PONs) will also require completely new approaches for optical modulation. The major challenge will be the design of modulation formats with simple generation and detection as well as excellent spectral efficiency, sensitivity and chromatic dispersion tolerance.

Novel coded modulation schemes address nonlinear impairment mitigation as well as mitigation of other detrimental effects such as polarization dependent loss and cycle slips due to excessive phase noise. For short reach applications, advanced modulation formats need to be designed specifically to match the properties of IM/DD systems.

REFERENCES

[1] E. Agrell et al., "Roadmap of optical communications", *Journal of Optics*, vol.18, 2016, doi:10.1088/2040-8978/18/6/063002.

[2] P. Sillard, "New fibers for ultra-high capacity transport," *Optical Fiber Technology*, vol.17, pp.495–502, 2011.

[3] R. J. Essiambre, G. Kramer, P. J. Winzer, G. J. Foschini e B. Goebel, "Capacity limits of optical fiber networks", *IEEE-Journal of Lightwave Technology*, vol.28, pp.662–701, 2010.

[4] P. J. Winzer, "High-spectral-efficiency optical modulation formats," *IEEE Journal of Lightwave Technology*, vol.30, pp.3824–3835, 2012.

[5] S. Randel et al., "*All-electronic flexibly programmable 864 Gb/s single-carrier PDM-64QAM,*" *Optical Fiber Communication (OFC) Conference*, San Francisco, USA, 2014.

[6] K. Schuh, E. Lach and B. Jungiger, "*100 Gbit/s ETDM transmission system based on electronic multiplexing transmitter and demultiplexing receiver,*" *32nd European Conference on Optical Communication (ECOC)*, Cannes, France, 2006.

[7] A. D. Shiner, M. Reimer, A. Borowiec, S. Oveis Gharan, J. Gaudette, P. Mehta, D. Charlton, K. Roberts and M. O'Sullivan, "Demonstration of an 8-dimensional modulation format with reduced interchannel nonlinearities in a polarization multiplexed coherent system", *Optics Express*, vol.22, pp.20366–20374, 2014.

Chapter 1

Optical Communications Systems

1.1 INTRODUCTION

The historical–technological evolution of the optical fiber communication systems can follow the path proposed by Chraplyvy [1, prg 2. History], having anyway in mind that the improvement of the system performance has been strictly related to the technological development of the optical fibers (Figure 1.1) and the optoelectronic devices, mainly semiconductor lasers (Figure 1.2), electro-optic modulators, optical receivers and optical amplifiers.

After the invention of the "optical waveguide" by Kao and Hockham in 1966 [2] and the demonstration of <20 dB/km optical waveguide loss by Corning Glass Works [3], the "Generation 0" (Table 1.1) of the optical fiber communication systems got started with multimode optical fibers and 850 nm optoelectronic technology, mainly based on GaAs/AlGaAs semiconductor material. In 1972, when Corning Glass announced the achievement of 4 dB/km loss in multimode optical fibers [4], the signal transmission on optical fibers started to be convenient as compared to the old coaxial cable transmission.

Even though the "Generation 1" (Table 1.2) of the optical fiber systems based on single-mode fibers (Figure 1.3) at 1300 nm and then at 1550 nm can be assumed to start in 1982 with the British Telecom field trial, during the preceding decade a considerable effort was directed to the technological development of the optical fibers, with the achievement of 0.2 dB/km attenuation at 1550 nm in 1979 [5] and of the semiconductor devices based on GaInAsP/InP technology (Figures 1.4 and 1.5) [6,7].

These research activities enabled the "Generation 1" to come into being with single-channel, Time-Division-Multiplexed (TDM) systems based on single-mode optical fibers at 1300 nm (II optical window), with 0.4 dB/km attenuation, and at 1550 nm (III optical window), with 0.2 dB/km attenuation [8]. In that context, with the increase of the bit rate from the Plesiochronous Digital Hierarchy (PDH) system (mainly, at 140 Mbit/s and 565 Mbit/s bit rates) to the Synchronous Digital Hierarchy (SDH) [9,10,11],

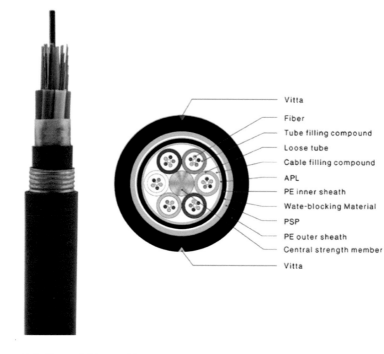

Figure 1.1 Optical fiber cable.

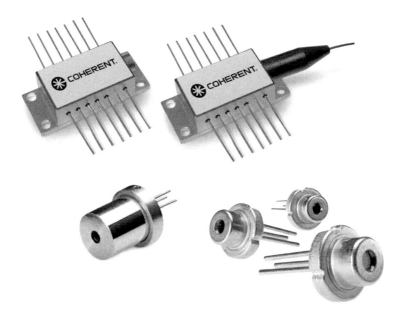

Figure 1.2 Semiconductor lasers.

Table 1.1 Generation 0

Kao Charles Kuen (Nobel Prize 2009) *Hockham George* *Invention of the "optical waveguide"*	*1966*
Optical fiber attenuation 20 dB/km ÷ 4 dB/km	
850 nm wavelength (I optical window) – GaAs/AlGaAs semiconductor technology	1966–1972
GaInAsP/InP semiconductor technology	1976–1977
0.2 dB/km optical fiber attenuation @ 1.55 µm (III optical window)	1979
Zero material dispersion in optical fiber @ 1.3 µm (II optical window)	1975–1977
Optical fiber communication system @ 140 Mbit/s	1977
Optical fiber communication system @ 565 Mbit/s	1978

Table 1.2 Generation 1

Single-mode optical fiber field trials @ 1.3 µm and 1.55 µm (British Telecom)	1982
Plesiochronous Digital Hierarchy (PDH) Recommendation ITU-T G.705 – Characteristics of Plesiochronous Digital Hierarchy (PDH) equipment functional block	1988 →
Standard single-mode optical fiber Recommendation ITU-T G.652 – Characteristics of a single- mode optical fiber and cable	1988 →
Dispersion-shifted single-mode optical fiber Recommendation ITU-T G.653 – Characteristics of a dispersion- shifted single-mode optical fiber and cable	1988 →
Synchronous Digital Hierarchy (SDH) Recommendation ITU-T G.783 – Characteristics of Synchronous Digital Hierarchy (SDH) equipment functional block	1990 →
SDH optical fiber transmission systems @ 2.5 Gbit/s and 10 Gbit/s	1991–1994
"Analog" binary optical coherent communication systems	1985–1995
"Analog" multilevel optical coherent communication systems	1990 →

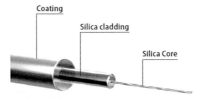

Figure 1.3 Single mode optical fiber.

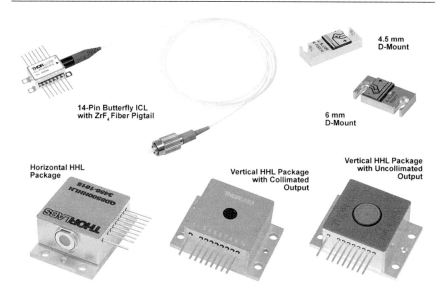

Figure 1.4 Further Semiconductor lasers.

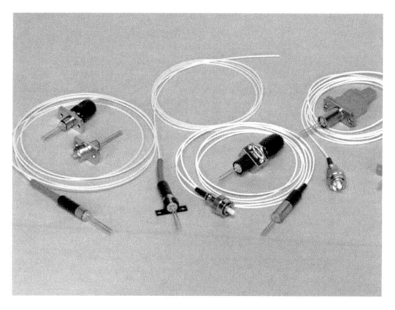

Figure 1.5 PIN Photodiodes.

at 2.5 Gbit/s, starting from 1991, and then at 10 Gbit/s (Figure 1.6), by the use of $LiNbO_3$ Mach–Zehnder electro-optic modulator (Figure 1.7) and high-performance PIN-FET optical receiver (Figure 1.8), the main problem to be solved was related to the high value of the chromatic dispersion (≈ 16–18 ps/km·nm) of the Standard Single-Mode Fiber (SSMF, ITU G.652).

Figure 1.6 10 Gbit/s optical receiver.

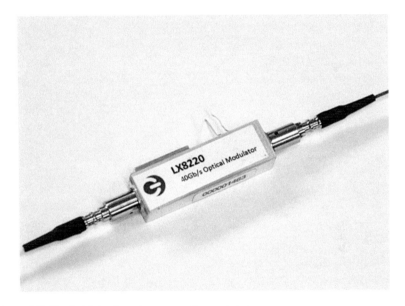

Figure 1.7 Mach–Zehnder electro-optic modulator.

That constraint was reduced by increasing the spectral purity of the optical sources (single-mode semiconductor laser) and definitively solved by the Dispersion-Shifted Fiber (DSF, ITU G.653), by which transmission at 1550 nm was possible without the issues due to the chromatic dispersion. As a consequence, the use of DSF enabled to realize 10 Gbit/s high-performance optical links. We could assume this goal, approximately around 1993, as the end of the "Generation 1" [12–19] and the start of the "Generation 2" (Table 1.3), which saw the transition from single-channel to multi-channel optical transmission by Wavelength Division Multiplexing (WDM) technique and the

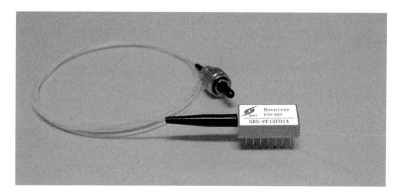

Figure 1.8 PIN-FET optical receiver.

Table 1.3 Generation 2

Wavelength Division Multiplexing (WDM) systems	1990
4λ WDM commercial system	1994
9 λ × 200 Mbit/s WDM system	1996
16 λ WDM commercial system	
Erbium-Doped Fiber Amplifier (EDFA) technology	1987–2002
4λ × 2.4 Gbit/s EDFA amplified (C optical band) WDM system	1990
EDFA in L optical band	1997 →
Nonlinear propagation in optical fiber (Kerr effect)	2001 →
Self-Phase Modulation (SPM), Four-Wave Mixing (FWM), Cross-Phase Modulation (XPM) nonlinear effects	2001–2010
Soliton fundamentals and applications	≈1986–2004
Non-zero dispersion (NZD) optical fiber	1993
Recommendation ITU-T G.655 – Characteristics of a non-zero dispersion-shifted single-mode optical fiber and cable	1996 →
Dispersion Management	1993
Dispersion Compensating Fiber (DCF)	1993
Raman distributed optical amplifier	≈1997–2002
Hybrid optical amplification EDFA + Raman	1999 →
Polarization Mode Dispersion (PMD)	1978 →
"Digital" multilevel optical coherent communication systems	2005 →
Recommendation ITU-T G.964.1 – Spectral grids for WDM applications: DWDM frequency grid	2012

use of Erbium-Doped Fiber Amplifier (EDFA, Figure 1.9) [20,21], proposed since 1987 [22,23].

In general terms, in addition to the increase in the device and system performance [24–37], the "Generation 2" marked the transition from the condition of linear propagation in optical fiber, characterized mainly by attenuation and chromatic dispersion issues, to the nonlinear propagation condition due to the Kerr effect [38] with the phenomena which it

Figure 1.9 Standard EDFA.

induces on the fiber propagation, namely Self-Phase Modulation (SPM) for single-channel transmission, Four-Wave Mixing (FWM) and Cross-Phase Modulation (XPM) for multi-channel transmission. While SPM can even give rise to positive effects like the Soliton transmission [38,39], XPM and FWM represent fundamental constraints for WDM transmission [30,34,40,41,42,43]. If XPM can be considered one of the main limiting effects for the communication capacity of a single-mode optical fiber [44,45], the intermodulation effects due to FWM destroy information supported by the WDM channels. It is well known that, because of FWM, DSF cannot support WDM transmission and the best way to tackle it is the reduction of the phase matching required by FWM. Thus, some chromatic dispersion that was a constraint in high bit rate single-channel transmission was the solution to support multi-channel transmission in WDM systems. Therefore, the condition to avoid the activation of the FWM led to the use of the Non-Zero Dispersion Fiber (NZDF), known as ITU G.655, proposed by Bell Laboratories around 1993.

The "Generation 2" was progressively stabilized during the successive decade [46–48] by the introduction of the Dispersion Management, since 1993 [49], the Dispersion Compensating Fiber (DCF) [50,51], the optical signal processing [52], the application of EDFA both in C band (1530–1565 nm) and L band (1565–1625 nm), the implementation of the "hybrid amplification" approach, by the combination of EDFA linear, concentrated amplification and nonlinear, distributed Raman amplification [53] (Figure 1.10), which has modified the paradigm of the signal amplification, at least in the optical domain (Figure 1.11).

Within this context, the quantum nature of the optical noise, so far neglected in the system design, represents the starting key point for the realization of very high-performance optical fiber communication systems [20,54]. In general terms, a new paradigm started to be followed in which, when possible, nonlinear effects are exploited to improve the system performance [55]. Another point to be faced and solved arose in the transition from 10 to 40 Gbit/s single-channel bit rate (Figure 1.12) that is the limitation due to the Polarization Mode Dispersion effect [30,34,36,46,48], related to the birefringence properties of the optical fiber.

The polarization degeneration of the fundamental mode gives rise to a linear but random dispersive effect, which can degrade the performance of 40 Gbit/s communication systems up to outage conditions. This problem was effectively solved by optical fibers with average values of the Differential

Figure 1.10 Raman module.

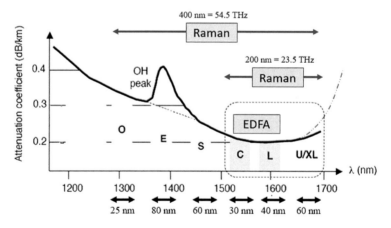

Figure 1.11 Bands allocation in the II and III optical windows.

Group Delay (DGD) between the two orthogonal polarization components in the range 0.1–0.2 ps/$\sqrt{km}$.

In general terms, it is possible to argue that the "Generation 2" concluded by the end of the first decade of the current century with commercial WDM systems supporting at least 80 channels at 40 Gbit/s, the demonstration of an overall transmission capacity up to 25 Tbit/s, and was stabilized in 2012 by the Recommendation ITU-T G.694.1 providing a frequency grid for Dense Wavelength Division Multiplexing (DWDM) systems that support channel spacings in the range from 12.5 GHz to 100 GHz and wider [56].

Figure 1.12 40 Gbit/s optical receiver.

Table 1.4 Generation 3

Optical Networking	
All-Optical Network (AON)	1992
Optical Transport Network (OTN)	
Recommendation ITU-T G.872 – Architecture of optical transport networks	1999 →
Reconfigurable Optical Add/Drop Multiplexer (ROADM)	2003 →
Optical Cross Connect (OXC)	2003 →
Broadband Optical Access Networks (FTTH standard)	2005 →
Spatial Multiplexing – Multicore Fiber (MCF)	2010 →
Optical Signal Processing	2010 →
"Digital" multidimensional optical coherent communication systems	2010 →

This last decade has seen the development of the "Generation 3" (Table 1.4) of the optical communications [57,58,91] in terms of networking (optical networks based on different configurations of Optical Add/Drop multiplexer, Optical Switcher and AWG - Arrayed Waveguide Grating, Figure 1.13) [59; 1, prg 5. Spatial multiplexing; 1 prg. 12. Long-haul networks], applications in the access network (PON – Passive Optical Network, Figures 1.14 and 1.15) [60; 1, prg. 13. Access networks], spatial multiplexing based on Multicore Fiber (MCF) [1, prg. 3. Optical fibers for next generation optical networks], optical signal processing [61; 1, prg. 9. Optical signal processing] and "digital" coherent optical systems based on multilevel modulation formats and DSP-based processing [62, 63; 1, prg. 6. Coherent transceivers; 1, prg. 7. Modulation formats] (Figures 1.16 and 1.17).

Figure 1.13 AWG Module.

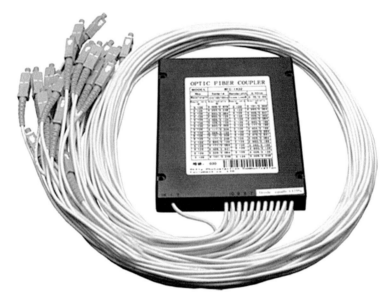

Figure 1.14 Star coupler.

In truth, the idea of realizing coherent optical systems started in 1980s, even during "Generation 1" when the serious problem of the wide spectral linewidth of the semiconductor lasers and its impact on the system performance were deeply investigated and solutions were found to solve the problems related to the State of Polarization (SOP) fluctuations due to the birefringence effect in optical fiber propagation [64,65,66]. During the decade 1985–1995, a lot of work was carried out relative to the conventional binary modulation formats (e.g., ASK, FSK, PSK), optical front ends (e.g., balanced and/or phase-diversity optical front ends), homodyne and heterodyne demodulation techniques (e.g., synchronous, envelope, differential, intradyne techniques). Besides the schemes of the optical front ends, the different technical solutions were mediated by RF and microwave technologies.

Figure 1.15 FTTH optical receiver.

Figure 1.16 100 Gbit/s optical receiver.

In the middle of 1990s, all this huge work led to define characteristics and performance of the different binary coherent optical communication schemes [67,68,69,70] that nowadays are collected under the denomination of "analog" coherent optical communication systems [48,71] so as to distinguish them from the present "digital" systems, based on conventional

Figure 1.17 100 Gbit/s optical receiver for DP-QPSK.

multilevel modulation formats and electronic processing by high-speed DSP [47,62,72,73], whose first demonstration dates back to 2005 [74].

In truth, at least as regards the application of multilevel modulation formats in coherent optical communication systems, this can be dated back to 1990, when multilevel modulation formats were investigated in non-conventional optical coherent systems, which were specifically designed to exploit the physical properties of the single-mode optical fiber.

The first proposal was based on a multilevel modulation technique in the polarization domain by the exploitation of the Stokes parameters formalism and the Poincaré sphere representation [75] and represents the natural evolution toward the multilevel signaling of the previous binary modulated system [76]. The second proposal exploited instead the properties of the phase and quadrature of the orthogonal components of the field polarization in the single-mode fiber and led to define a multilevel modulation format in a four-dimensional (4-quadratures of the electromagnetic field in the optical fiber) physical space [77,78]. Even though a similar approach was analyzed from the point of view of "four-dimensional Coding" [79], the proposed system configuration opened a novel perspective for multilevel signaling in optical fiber since information is represented in a four-dimensional space that represents the maximum dimensionality of the physical space used for electromagnetic communications for a given power value that is within a spherical symmetry. This "reference" system has been then deeply analyzed in terms of power efficiency [80,81,82,83] and in combination with coding techniques [84,85,86]. Its properties have been derived from the physical point of view starting from the fundamental quantum mechanical boson commutation relations and relating the system performance to the possible physical and non-physical degrees of freedom in four-dimensional space [87]. A version for satellite communications was proposed as well [88].

A general result can be deduced since from the 4-Quadrature Signaling, all the binary and multilevel, non-coded, constant power modulation formats designed in two- and three-dimensional spaces can be derived through the formalism proposed by Karlsson [87].

In particular, as demonstrated in the following Chapter 3, the three-dimensional modulation format in the Stokes space [75] can be derived by the 4-Quadrature scheme by suitable "one-dimensional space compression" [89,90], as well as the classical multilevel phase modulations by "two-dimensional space compression".

Therefore, the formalism of the quadratic forms with the constraint of constant power can be used to represent all the modulation formats by space rotations and compressions. As demonstrated by Karlsson, of the possible six degrees of freedom of a topological four-dimensional space, four are related to physical magnitudes while two have only mathematical meaning. This means that four dimensions can be directly exploited for information transmission by physical parameters, the remaining two can be used in "virtual" manner by electronic signal processing based on high bit rate DSP.

Among the points to be clarified, the following ones seem to be of particular interest:

- For a given constellation in four-dimensional space, which are the corresponding constellations in the sub-dimension spaces (space compression and rotations)? How can the double or triple points in the sub-spaces be exploited from the information point of view?
- Is it possible to define a simple algorithm that enables to calculate the performance of a system in three- or two-dimensional space derived from a constellation assembled within the four-dimensional sphere?

The particular structure of the "reference" 4-Quadrature system opens even new perspectives as regards the exploitation of constellation points not for information transmission but for "signal" protection by "coding" techniques based on a different use of the constellation points and the effect due to the passage from the four-dimensional space to the sub-spaces. This operation could modify the information content for a given constellation so as to have an "entropic" effect but, at the same time, to provide a sort of "natural coding" without band expense.

It is reasonable to think that the "Generation 3" can be closed when the above and other questions can have answer.

REFERENCES

[1] E. Agrell et al., "Roadmap of optical communications", *Journal of Optics*, vol.18, 2016, doi: 10.1088/2040-8978/18/6/063002.

[2] K.C. Kao and A. Hockham, "Dielectric-fibre surface waveguides for optical frequencies", *Proceedings of the Institution of Electrical Engineers*, vol.113, n°7, pp.1151–1158, July 1966.

[3] F.P. Kapron, D.B. Keck and R.D. Maurer, "Radiation losses in glass optical waveguides", *Applied Physics Letters*, vol.17, n°10, pp.423–425, November 1970, https://doi.org/10.1063/1.1653255.

[4] R.D. Maurer, "First European electro-optics market and technology", September 12–15, 1972, Geneva (CH).

[5] T. Miyashita, T. Miya and M. Nakahara, "*An ultimate low loss single mode fiber at 1.55 µm*", OFC, Optical Fiber Communication, March 6–8, 1979, Washington, DC.

[6] J. J. Hsieh, "Room temperature operation of GaInAsP/InP double heterostructure diode laser emitting at 1.1 µm", *Applied Physics Letters*, vol.28, n°5, pp.283–285, March 1976.

[7] C. C. Shen, J. J. Hsieh and T. A. Lind, "1500 h continuous cw operation of double-heterostructure GaInAsP/InP lasers", *Applied Physics Letters*, vol.30, n°7, pp.353–354, April 1977, https://doi.org/10.1063/1.89397

[8] S. E. Miller, I. Kaminow, *Optical Fiber Telecommunications II*, Academic Press, New York, 1988.

[9] R. Ballart and Y.-C. Ching, "SONET: Now it's the standard optical network", *IEEE Communications Magazine*, Vol.27, n°3, March 1989, https://doi.org/10.1109/35.20262

[10] *IEEE Communications Magazine*, vol.28, n°8, August 1990.

[11] C.A. Siller, M. Shafi, "*SONET/SDH: A Sourcebook of Synchronous Networking*", IEEE Press, New York, 1996.

[12] H.B. Killen, *Fiber-Optic Communications*, Prentice Hall, Englewood Cliffs, NJ, 1991.

[13] G. E. Keiser, *Optical Fiber Communications*, 2nd ed., McGraw-Hill, New York, 1991.

[14] T. Li (Editor), *Topics in Lightwave Transmission Systems*, Academic Press, Boston, 1991.

[15] G. P. Agrawal, *Fiber-Optic Communication Systems*, Wiley, New York, 1992.

[16] J. C. Palais, *Fiber-Optic Communications*, 3rd ed., Prentice Hall, Englewood Cliffs, NJ, 1992.

[17] J. M. Senior, *Optical Fiber Communications*, 2nd ed., Prentice Hall, New York, 1992.

[18] J. Gowar, *Optical Communication Systems*, 2nd ed., Prentice Hall, Englewood Cliffs, NJ, 1993.

[19] L. D. Green, *Fiber Optic Communications*, CRC Press, Boca Raton, 1993.

[20] E. Desurvire, *Erbium-Doped Fiber Amplifiers – Principles and Applications*, Wiley-Interscience, Boca Raton, NJ, 2002.

[21] E. Desurvire, D. Bayart, B. Desthieux, S. Bigo, *Erbium-Doped Fiber Amplifiers – Device and System Developments*, Wiley-Interscience, New York, 2002.

[22] R. J. Mears et al., "Low-noise erbium-doped fibre amplifier operating at 1.54 µm", *Electronics Letters*, vol.23, n°19, pp.1026–1028, September 1987, doi: 10.1049/el:19870719.

[23] E. Desurvire, J. R. Simpson and P. C. Becker, "High-gain erbium-doped traveling-wave fiber amplifier", *Optics letters*, vol.12, n°11, pp.888–890, November 1987.

[24] G. Einarrson, *Principles of Lightwave Communication Systems*, Wiley, New York, 1995.

[25] J. Franz, *Optical Communication Systems*, Narosa Publishing House, New Delhi, 1996.

[26] M. M.-K. Liu, *Principles and Applications of Optical Communications*, Irwin, Chicago, 1996.

[27] L. Kazovsky, S. Benedetto and A. E. Willner, *Optical Fiber Communication Systems*, Artech House, Boston, 1996.

[28] R. Papannareddy, *Introduction to Lightwave Communication Systems*, Artech House, Boston, 1997.

[29] G. P. Agrawal, *Fiber-Optic Communication Systems*, 2nd ed., Wiley, New York, 1997.

[30] I. P. Kaminow, T. L. Koch (Editors), *Optical Fiber Telecommunications III A*, Academic Press, San Diego, CA, 1997.

[31] I. P. Kaminow, T. L. Koch (Editors), *Optical Fiber Telecommunications III B*, Academic Press, San Diego, CA, 1997.

[32] G. Keiser, *Optical Fiber Communications*, 3rd ed., McGraw-Hill, New York,, 2000.

[33] I. P. Kaminow, Tingye Li (Editors), *Optical Fiber Telecommunications IV A*, Academic Press, San Diego, CA, 2002.

[34] I. P. Kaminow, Tingye Li (Editors), *Optical Fiber Telecommunications IV B*, Academic Press, San Diego, CA, 2002.

[35] G. P. Agrawal, *Lightwave Technology – Components and Devices*, Wiley-Interscience, Hoboken, NJ, 2004.

[36] G. P. Agrawal, *Lightwave Technology – Telecommunication Systems*, Wiley-Interscience, Hoboken, NJ, 2005.

[37] R. Ramaswami, K. N. Sivarajan, *Optical Networks – A Practical Perspective*, 2nd ed., Morgan Kaufmann Publishers, 2002.

[38] G. P. Agrawal, *Nonlinear Fiber Optics*, 3rd ed., Academic Press, Boston, 2001.

[39] L. F. Mollenauer and J. P. Gordon, *Solitons in Optical Fibers – Fundamentals and Applications*, Elsevier Academic Press, Boston, 2006.

[40] S. Betti, M. Giaconi, "Analysis of the cross-phase modulation effect in WDM optical systems", *IEEE-Photonics Technology Letters*, vol.13, n°1, pp.43–45, January 2001.

[41] S. Betti, M. Giaconi, "Effect of the cross-phase modulation on WDM optical systems: Analysis of fiber propagation", *IEEE – Photonics Technology Letters*, vol.13, n°4, pp.305–307, April 2001.

[42] S. Betti, M. Giaconi, M. Nardini, "Effect of Four-wave mixing on WDM optical systems: A statistical analysis", *IEEE-Photonics Technology Letters*, vol.15, n°8, pp.1079–1081, August 2003.

[43] L. G. L. Wegener et al., "The effect of propagation nonlinearities on the information capacity of WDM optical fiber systems: Cross-phase modulation and four-wave mixing", *Physica D: Nonlinear Phenomena*, vol.189, n°1–2, pp.81–99, 2004.

[44] P. P. Mitra and J. B. Stark, "Nonlinear limits to the information capacity of optical fibre communications", *Nature*, vol.411, pp.1027–1030, 2001.

[45] A. D. Ellis, J. Zhao and D. Cotter, "Approaching the non-linear Shannon limit", *IEEE – Journal of Lightwave Technology*, vol.28, n°4, pp.423–433, February 2010.

[46] I. P. Kaminow, Tingye Li, A. E. Willner (Editors), *Optical Fiber Telecommunications V A: Components and Subsystems*, Academic Press, London, 2008.

[47] I. P. Kaminow, Tingye Li, A. E. Willner (Editors), *Optical Fiber Telecommunications V B: Systems and Networks*, Academic Press, London, 2008.

[48] G. P. Agrawal, *Fiber-Optic Communication Systems*, 4th ed., Wiley, Hoboken, NJ, 2010.

[49] A. R. Chraplyvy et al., "8×10 Gb/s transmission through 280 km of dispersion-managed fiber", *IEEE-Photonics Technology Letters*, vol.5, n°10, pp.1233–1235, October 1993.

[50] A. M. Vengsarkar, A.E. Miller and W. A. Reed, "*Highly efficient single-mode fiber for broadband dispersion compensation*", Optical Fiber Communication Conference, February 1993, San Jose, CA, PD-13.

[51] S. Ramachandran, *Fiber Based Dispersion Compensation*, vol.5, Springer, New York, 2007.

[52] B. J. Eggleton et al., "Photonic chip based ultrafast optical processing based on high nonlinearity dispersion engineered chalcogenide waveguides", *Laser & Photonics Reviews*, vol.6, n°1, pp.97–114, 2012.

[53] Mohammed N. Islam (Editor), *Raman Amplifiers for Telecommunications 2 – Sub-Systems and Systems*, Springer-Verlag, New York, 2004.

[54] C. Headley, G. P. Agrawal (Editors), *Raman Amplification in Fiber Optical Communication Systems*, Elsevier-Academic Press, Amsterdam, 2005.

[55] G. P. Agrawal, *Applications of Nonlinear Fiber Optics*, 2nd ed., Elsevier-Academic Press, Amsterdam, 2008.

[56] ITU-T G.694.1 (02/2012), SERIES G: Transmission systems and media, digital systems and networks – Transmission media and optical systems characteristics – Characteristics of optical systems, "Spectral grids for WDM applications: DWDM frequency grid".

[57] I. P. Kaminow, Tingye Li, A. E. Willner (Editors), *Optical Fiber Telecommunications VI A: Components and Subsystems*, Academic Press, Amsterdam, 2013.

[58] I. P. Kaminow, Tingye Li, A. E. Willner (Editors), *Optical Fiber Telecommunications VI B: Systems and Networks*, Academic Press, Amsterdam, 2013.

[59] R.-J. Essiambre and R. W. Tkach, "Capacity trends and limits of optical communication networks", *Proceedings of the IEEE*, vol.100, n°5, pp.1035–1055, May 2012.

[60] L. G. Kazovsky et al., *Broadband Optical Access Networks*, Wiley, Hoboken, NJ, 2011.

[61] S. Wabnitz and B. J. Eggleton (Editors), *All-Optical Signal Processing: Data Communication and Storage Applications*, Springer, Cham, Switzerland, 2015.

[62] K. Kikuchi, "Fundamentals of coherent optical fiber communications", *IEEE – Journal of Lightwave Technology*, vol.34, n°1, pp.157–179, January 2016.

[63] P. J. Winzer, "High-spectral-efficiency optical modulation formats", *IEEE-Journal of Lightwave Technology*, vol.30, n°24, pp.3824–3835, December 2012.

[64] T. Okoshi and K. Kikuchi, *Coherent optical Communication Systems*, KTK Scientific Publisher, Tokyo, Japan, 1988.

[65] R. A. Linke and A. H. Gnauck, "High-capacity coherent lightwave systems", *IEEE – Journal of Lightwave Technology*, vol.6, n°11, pp.1750–1769, November 1988.

[66] P. S. Henry, *Coherent Lightwave Communications*, IEEE Press, New York, 1990.

[67] P. W. Hooijmans, *Coherent Optical System Design*, Wiley, Chichester, 1994.

[68] S. Ryu, *Coherent Lightwave Communication Systems*, Artech House, Boston, 1995.

[69] S. Betti, G. De Marchis, E. Iannone, *Coherent Optical Communications Systems*, Wiley, New York, 1995.

[70] M. Cvijetic, *Coherent and Nonlinear Lightwave Communications*, Artech House, Boston, 1996.

[71] Keang-Po Ho, *Phase-Modulated Optical Communication Systems*, Springer, New York, 2005.

[72] E. Ip et al., "Coherent detection in optical fiber systems", *Optics Express*, vol.16, n°2, pp.753–791, January 2008.

[73] Le Nguyen Binh, *Optical Modulation – Advanced Techniques and Applications in Transmission Systems and Networks*, CRC Press, Boston, 2018.

[74] S. Tsukamoto, D.-S. Ly-Gagnon, K. Katoh and K. Kikuchi, "*Coherent demodulation of 40-Gbit/s polarization-multiplexed QPSK signals with 16-GHz spacing after 200-km transmission*", *Optical Fiber Communication Conference*, March 6–11, 2005, Anaheim, CA, PDP-29.

[75] S. Betti, F. Curti, G. De Marchis, E. Iannone, "Multilevel coherent optical system based on Stokes parameters modulation", *IEEE – Journal of Lightwave Technology*, vol.8, n°7, pp.1127–1136, July 1990.

[76] S. Betti, F. Curti, G. De Marchis, E. Iannone, "Phase noise and polarization state insensitive optical coherent systems", *IEEE – Journal of Lightwave Technology*, vol.8, n°5, pp.756–767, May 1990.

[77] S. Betti, F. Curti, G. De Marchis, E. Iannone, "Exploiting fibre optics transmission capacity: 4-Quadrature multilevel signalling", *Electronics Letters*, vol.26, n°14, pp.992–993, July 1990.

[78] S. Betti, F. Curti, G. De Marchis, E. Iannone, "A novel multilevel coherent optical system: 4-Quadrature signaling", *IEEE – Journal of Lightwave Technology*, vol.9, n°4, pp.514–523, April 1991.

[79] G. R. Welti and Jhong S. Lee, "Digital transmission with coherent four-dimensional modulation", *IEEE Transactions on Information Theory*, vol.IT-20, n°4, pp.497–502, July 1974.

[80] M. Karlsson and E. Agrell, "Which is the most power-efficient modulation format in optical links?", *Optics Express*, vol.17, n°13, pp.10814–10819, June 2009.

[81] E. Agrell and M. Karlsson, "Power-efficient modulation formats in coherent transmission systems", *IEEE – Journal of Lightwave Technology*, vol.27, n°22, pp.5115–5126, November 2009.

[82] M. Karlsson, E. Agrell, "*Four-dimensional optimized constellations for coherent optical transmission systems*", *ECOC 2010*, September 19–23, 2010, Torino, Italy, We.8.C.3.

[83] M. Karlsson and E. Agrell, "*Spectrally efficient four-dimensional modulation*", *Optical Fiber Communication (OFC) Conference*, March 4–8, 2012, Los Angeles, CA, OTu2C.1, https://doi.org/10.1364/OFC.2012.OTu2C.1.

[84] L. Beygi, E. Agrell, J. M. Kahn and M. Karlsson, "Coded modulation for fiber-optic networks – Toward better tradeoff between signal processing complexity and optical transparent reach", *IEEE – Signal Processing Magazine*, vol.31, n°2, pp.93–103, March 2014.

[85] M. Karlsson and E. Agrell, "*Multidimensional modulation and coding*", *Optical Fiber Communications Conference*, March 20–22, 2016, Anaheim, CA, M3A.1, https://doi.org/10.1364/OFC.2016.M3A.1.

[86] M. Karlsson and E. Agrell, "Multidimensional modulation and coding in optical transport", *IEEE – Journal of Lightwave Technology*, vol.35, n°4, pp.876–884, February 2017.

[87] M. Karlsson, "Four-dimensional rotations in coherent optical communications", *IEEE – Journal of Lightwave Technology*, vol.32, n°6, pp.1246–1257, March 2014.

[88] L. Arend et al., "Four-dimensional signalling schemes – Application to satellite communications", arXiv:1511.04557 [cs.IT], November 2015.

[89] G. G. Rutigliano, S. Betti, P. Perrone, *"Representations of optical fiber communications in four and three dimensional spaces"*, *20th Italian National Conference on Photonic Technologies (Fotonica 2018)*, May 23–25, 2018, Lecce, Italy, IET, doi: 10.1049/cp.2018.1624.

[90] G. G. Rutigliano, S. Betti, P. Perrone, "Representations of optical fibre communications in three- and four-dimensional spaces", *IET Communications*, vol.13, n°20, pp.3558–3564, December 2019, DOI: 10.1049/iet-com.2019.0277.

[91] A. E. Willner, *Optical Fiber Telecommunications VII*, Academic Press, 2019.

Chapter 2

Polarization in Optical Fiber

2.1 OPTICAL FIBERS

The phenomenon of total internal reflection, responsible for guiding of light in optical fibers, has been known since 1854 [1]. Although glass fibers were made in the 1920s [2, 3, 4], their use became practical only in the 1950s with the use of a cladding layer. Before 1970, optical fibers were used mainly for medical imaging over short distances [5]. Their use for communication purposes was considered impractical because of high losses (~1000 dB/km). However, the situation changed drastically in 1970 when the loss of optical fibers was reduced to below 20 dB/km [6]. In 1979, further progress led to a loss of only 0.2 dB/km near the 1.55-μm spectral region [7]. The availability of low-loss fibers led to a revolution in the field of lightwave technology and started the era of fiber-optic communications (Figure 2.1).

The propagation of electromagnetic field in optical fiber is governed by Maxwell's equations expressed in the following form:

$$\nabla \times E = -\partial B / \partial t, \tag{2.1}$$

$$\nabla \times H = \partial D / \partial t, \tag{2.2}$$

$$\nabla \cdot D = 0, \tag{2.3}$$

$$\nabla \cdot B = 0, \tag{2.4}$$

where E and H are the electric and magnetic field vectors, respectively, and D and B are the corresponding flux densities. To study the wave propagation in optical fibers, it is possible to refer to the wave equation:

$$\nabla^2 E + \varepsilon k_0^2 E = 0 \tag{2.5}$$

where ε is the dielectric tensor and $k_0 = 2\pi/\lambda_0$ is the propagation constant of free space, with λ_0 the free space wavelength.

Figure 2.1 Optical fiber cable.

An optical mode refers to a specific solution of the wave Equation (2.5) that satisfies the appropriate boundary conditions and has the property that its spatial distribution does not change with propagation. The fiber modes can be classified as guided modes, leaky modes, and radiation modes [8]. Signal transmission in fiber-optic communication systems takes place through the guided modes only. We will analyze only Standard Single-Mode Fibers (SSMFs), which support only the so-called fundamental mode of the fiber (HE_{11}). The fiber is designed so that all higher-order modes are cut off at the operating wavelength. The fundamental mode has no cut-off and it is always supported by a fiber. A single-mode fiber actually supports two orthogonally polarized modes that are degenerate and have the same mode index:

$$E(x,y,z) = \left[a_x(z) E_x(x,y) e^{-i\beta_x(z)} + a_y(z) E_y(x,y) e^{-i\beta_y(z)} \right], \qquad (2.6)$$

where the attenuation term has been neglected for simplicity, z is the propagation axis, $E_x(x, y)$ and $E_y(x, y)$ are the electric fields associated to the two orthogonally polarized modes H_{11}^x and H_{11}^y, β_x and β_y are the propagation constants and $a_x(z)$, $a_y(z)$ the complex functions that may be obtained by requiring that the expression (2.6) is a solution of the Equations (2.1)–(2.4). Under the hypothesis of "weak" perturbations, maintaining unchanged anyway the modal structure, with the further assumption that $\beta_x \cong \beta_y$, it can be shown that the functions $a_x(z)$, $a_y(z)$ satisfy the following coupled equations:

$$\frac{da_x(z)}{dz} = -i\left[k_{xx} a_x(z) + k_{xy} a_y(z) \right] \qquad (2.7)$$

$$\frac{da_y(z)}{dz} = -i\left[k_{yx} a_x(z) + k_{yy} a_y(z) \right], \qquad (2.8)$$

where the k_{ij} terms are the coupled coefficients.

The degenerate nature of the orthogonally polarized modes holds only for an ideal single-mode fiber with a perfectly cylindrical core of uniform diameter. Real fibers exhibit considerable variation in the shape of their core along the fiber length. They may also experience non-uniform stress so that the cylindrical symmetry of the fiber is broken.

All polarization effects in SMF (Single-Mode Fiber) are a direct consequence of the accidental loss of degeneracy for the polarization modes.

2.2 POLARIZATION AND RADIATION FIELD

The electromagnetic wave polarization represents how the orientation of the electric field components in a plane, which is perpendicular to propagation direction, varies. The general polarization state of the wave is the elliptical one. The special cases are the circular and linear states.

Consider an electromagnetic wave which is described by the electric field intensity vector $\mathbf{E}$ and which propagates in the direction of z axis. The wave may be represented as a superposition of two partial waves with mutually orthogonal linear polarizations and with the same frequency:

$$E = E_x + E_y \tag{2.9}$$

where $\mathbf{E}_x$ and $\mathbf{E}_y$ are the complex vectors of electrical field intensity, aligned along x axis and y axis directions, respectively. These two vectors may be assigned to two degenerate modes of the single-mode fiber, which is a dielectric circular waveguide:

$$E_x(z,t) = E_{0,x} e^{-i(\omega t - kz - \phi_x)}, \tag{2.10}$$

$$E_y(z,t) = E_{0,y} e^{-i(\acute{E}t - kz - \phi_y)}, \tag{2.11}$$

where $E_{0,x}$ and $E_{0,y}$ are the wave amplitudes, ω is the angular frequency of the waves, t the time, ϕ_x, ϕ_y are the phases of the wave and k is the medium propagation constant.

By means of an opportune choice of the four parameters $E_{0,x}$, $E_{0,y}$, ϕ_x and ϕ_y, it is possible to describe any State Of Polarization (SOP) of coherent light.

If the four amplitude and phase parameters of the field do not assume special values, the wave describes, in the transverse plane with respect to the propagation direction, a "polarization ellipse" that can be represented in the following way:

$$\frac{E_x^2}{E_{0,x}^2} + \frac{E_y^2}{E_{0,y}^2} - 2\frac{E_x E_y}{E_{0,x} E_{0,y}} \cos(\phi_y - \phi_x) = \sin^2(\phi_y - \phi_x). \tag{2.12}$$

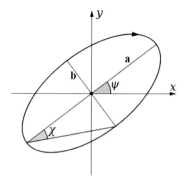

Figure 2.2 Polarization ellipse with the orientation angle ψ and the ellipticity χ.

Denoting with ψ the angle between the main axis of the ellipse and the x axis, it can be shown that:

$$\tan 2\psi = \frac{2E_{0,x}E_{0,y}\cos(\phi_x - \phi_y)}{E_{0,x}^2 - E_{0,y}^2}. \qquad (2.13)$$

Another important parameter is the polarization ellipse ellipticity (Figure 2.2) defined as:

$$\tan 2\chi = \pm\frac{b}{a}, \qquad (2.14)$$

and it can be shown that, in terms of the field parameters, the following relationship holds:

$$\sin 2\chi = \frac{2E_{0,x}E_{0,y}\sin(\phi_x - \phi_y)}{E_{0,x}^2 + E_{0,y}^2}. \qquad (2.15)$$

A lightwave propagates in homogenous isotropic medium with a constant velocity, independently on the propagation direction. The propagation velocity v is:

$$v = c/n, \qquad (2.16)$$

where c is the wave velocity in vacuum and n the medium refractive index. However, a medium may be of anisotropic character. This means that the propagation velocity depends on the propagation direction. This effect is observed in birefringent materials. When a lightwave propagates in such type of material as described above, a phase shift between its orthogonal components may occur due to different propagation velocities of the components. The total birefringence $\Delta\beta$ leads to a periodic power exchange

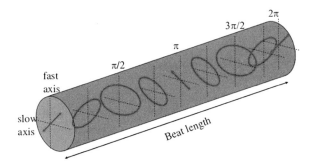

Figure 2.3 State of polarization in a birefringent fiber over one beat length.

between the two polarization components. The period, referred to as the beat length L_B, is given by:

$$L_B = \lambda / \Delta\beta. \tag{2.17}$$

Typically, $\Delta\beta \approx 10^{-7}$ and $L_B \approx 10$ m for $\lambda \approx 1$ μm. From a physical point of view, the state of polarization changes along the fiber length in a periodic manner over the length L_B. Figure 2.3 shows schematically such a periodic change in the state of polarization for a fiber of constant birefringence and an input beam linearly polarized at 45° with respect to the *slow* and *fast* axes. The *fast* axis in this figure corresponds to the axis along which the mode index is smaller. The other axis is called the *slow* axis.

In conventional single-mode fibers, birefringence is not constant along the fiber. Moreover, different frequency components of a pulse acquire different polarization states, resulting in pulse broadening. This phenomenon is called Polarization Mode Dispersion (PMD) and becomes a limiting factor for optical communication systems operating at high bit rates.

2.2.1 Jones Representation

The description of the polarizing behavior of the optical field in terms of amplitudes was one of the first great successes of the wave theory of light. The solution of the wave Equation (2.5), in terms of transverse components, leads to elliptically polarized light and its degenerate linear and circular forms. On the basis of the amplitude results, many concepts could be understood (e.g., Young's interference experiment, circularly polarized light). However, even using the amplitude formulation, numerous problems become difficult to treat, such as the propagation of the field through several polarizing components. To facilitate the treatment of complicated polarization problems at the amplitude level, R. Clark Jones, in the early 1940s, developed a matrix calculus for treating these problems, commonly called the Jones matrix calculus. The Jones calculus involves complex quantities contained in 2×1 column matrices (the Jones vector) and 2×2 matrices (the Jones matrices). In general, it is used for fully polarized light.

If the propagator term (ωt – kz) is suppressed from (2.10) and (2.11), the two field components can be written as:

$$E_x = E_{0,x}\, e^{i\phi_x}, \tag{2.18}$$

$$E_y = E_{0,y}\, e^{i\phi_y}. \tag{2.19}$$

They can be arranged in a 2×1 column matrix **E**:

$$E = \begin{pmatrix} E_x \\ E_y \end{pmatrix} = \begin{pmatrix} E_{0,x}\, e^{i\phi_x} \\ E_{0,y}\, e^{i\phi_y} \end{pmatrix} \tag{2.20}$$

called the Jones column matrix or, simply, the Jones vector. In the Jones vector (2.20), the maximum amplitudes $E_{0,x}$ and $E_{0,y}$ are real quantities. The presence of the exponent with imaginary arguments causes E_x and E_y to be complex quantities. The total intensity I of the optical field is given by:

$$I = E_x E_x^* + E_y E_y^* = \begin{pmatrix} E_x^* & E_y^* \end{pmatrix} \begin{pmatrix} E_x \\ E_y \end{pmatrix}. \tag{2.21}$$

Table 2.1 shows the Jones vectors of the fundamental states of polarization.

Table 2.1 Jones vectors of the fundamental SOPs

SOP	Jones Vector
Linear Horizontally Polarized light (LHP)	$\begin{pmatrix} 1 \\ 0 \end{pmatrix}$
Linear Vertically Polarized light (LVP)	$\begin{pmatrix} 0 \\ 1 \end{pmatrix}$
Linear +45° polarized light (L+45)	$\dfrac{1}{\sqrt{2}}\begin{pmatrix} 1 \\ 1 \end{pmatrix}$
Linear −45° polarized light (L-45)	$\dfrac{1}{\sqrt{2}}\begin{pmatrix} 1 \\ -1 \end{pmatrix}$
Right-hand Circularly Polarized light (RCP)	$\dfrac{1}{\sqrt{2}}\begin{pmatrix} 1 \\ i \end{pmatrix}$
Left-hand Circularly Polarized light (LCP)	$\dfrac{1}{\sqrt{2}}\begin{pmatrix} 1 \\ -i \end{pmatrix}$

In linear propagation conditions, in correspondence to a distance z from the input of the fiber, the electromagnetic field E (z, ω) can be expressed by the transfer matrix T (the Jones matrix), according to the relationship:

$$E(z,\omega) = T(z,\omega)E_0 \tag{2.22}$$

where E_0 is the initial field and T (z, ω) is:

$$T(z,\omega) = e^{(-\alpha/2 + i\beta)z} \, U(z,\omega). \tag{2.23}$$

The terms α and β consider, respectively, the attenuation and the chromatic dispersion of the fiber and U (z, ω) is a unitary matrix with complex components:

$$U = \begin{pmatrix} u_1 & u_2 \\ -u_2^* & u_1^* \end{pmatrix} = \begin{pmatrix} \epsilon e^{i\zeta_1} & \sqrt{1-\epsilon^2}\, e^{i\zeta_2} \\ -\sqrt{1-\epsilon^2}\, e^{-i\zeta_2} & \epsilon e^{-i\zeta_1} \end{pmatrix} \tag{2.24}$$

with $|u_1|^2 + |u_2|^2 = 1$, $U^T = U^{-1}$ and ϵ, ζ_1, ζ_2 are three independent real parameters [9, 10].

2.2.2 Stokes Representation

George Gabriel Stokes introduced the Stokes parameters for the first time in 1852 [11]. They exactly define a state of polarization or, in other words, a state of polarization is perfectly known when the relative Stokes parameters are known. The four parameters $E_{0,x}$, $E_{0,y}$, ϕ_x and ϕ_y (introduced in paragraph 3.1) identify the Stokes parameters in the following way [11]:

$$S_0 = \left[E_{0,x}^2 + E_{0,y}^2 \right] \tag{2.25}$$

$$S_1 = \left[E_{0,x}^2 - E_{0,y}^2 \right], \tag{2.26}$$

$$S_2 = \left[2E_{0,x}E_{0,y} \cos(\phi_y - \phi_x) \right] \tag{2.27}$$

$$S_3 = \left[2E_{0,x}E_{0,y} \sin(\phi_y - \phi_x) \right]. \tag{2.28}$$

It is possible to derive from (2.25)–(2.28) the following relationship:

$$S_0^2 = S_1^2 + S_2^2 + S_3^2. \tag{2.29}$$

Expression (2.29), which links the four Stokes parameters, introduces the definition of the Polarization Degree (P_D):

$$P_D = \frac{I_{polarized}}{I_{total}} = \frac{\sqrt{S_1^2 + S_2^2 + S_3^2}}{S_0}, \tag{2.30}$$

that represents the ratio between the intensity of the light which, in the observation time, can be measured in its state of polarization (polarized light $I_{polarized}$) and the overall intensity of the light (I_{total}). The degree of polarization of the light can assume values from a maximum of 1 to a minimum of zero for a completely unpolarized light.

The parameter S_0 represents the total intensity of the polarized optical beam. The S_1 parameter represents the difference between the two polarization components (horizontal and vertical). The parameters S_2 and S_3 represent the difference between the polarization components measured at ±45° (the S_2 parameter) and the difference between the left and right circular polarizations (the S_3 parameter). The range of possible values for S_1, S_2 and S_3 is $[-1,1]$. From (2.13) and (2.15), it is possible to derive:

$$\tan 2\psi = \frac{S_2}{S_1}, \tag{2.31}$$

$$\sin 2\chi = \frac{S_3}{S_0}, \tag{2.32}$$

from which, making use also of the relationship (2.29), the three Stokes parameters can be represented as functions of S_0 and of the two angles χ and ψ:

$$S_1 = S_0 \cos 2\chi \cos 2\psi, \tag{2.33}$$

$$S_2 = S_0 \cos 2\chi \sin 2\psi, \tag{2.34}$$

$$S_3 = S_0 \sin 2\chi. \tag{2.35}$$

These relationships are the same as those that bind the Cartesian coordinates to the spherical coordinates of a sphere of radius S_0 with the condition that the angles on the sphere are double with respect to those of the ellipse in Figure 2.2. Usually, it is preferred to normalize the Stokes parameters to the intensity value S_0; in this way, it is possible to refer to a sphere of unit radius. The formalism of the Stokes parameters is completely capable of describing any polarization state ranging from completely polarized light to completely unpolarized light. In addition, this formalism can be used to describe the superposition of several polarized beams, provided that there is no amplitude or phase relation between them (beams incoherent with respect to each other).

This situation arises when optical beams are emitted from several independent sources and they are then superposed.

Henri Poincaré, a famous nineteenth-century French mathematician and physicist, discovered, around 1890, that the polarization ellipse of Figure 2.2 could be represented on a complex plane. Further, he discovered that this plane could be projected onto a sphere in exactly the same manner as the stereographic projection. In effect, he reversed the problem of classical antiquity, which was to project a sphere onto a plane. The sphere that Poincaré devised is extremely useful for dealing with polarized light problems and, appropriately, it is called the *Poincaré sphere*. In 1892, Poincaré introduced his sphere [12] in which 2ψ is the angle on the equatorial plane and 2χ is the altitude angle on the equator (Figure 2.4).

Figure 2.5 shows the representation onto the Poincaré sphere of the fundamental SOPs described in paragraph 2.2.1.

There is therefore a bi-univocal correspondence between points of the Poincaré sphere and the states of polarization, either if they are represented

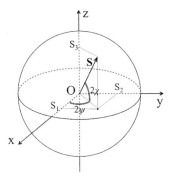

Figure 2.4 Poincaré sphere.

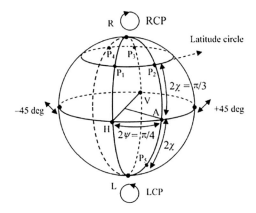

Figure 2.5 Fundamental SOPs represented on the Poincaré sphere.

with the Stokes parameters or with the Jones parameters. Called x, y and z the axes with respect to which the sphere is traced, it is possible to identify specific points on the sphere itself. The equatorial plane of the sphere ($2\chi = 0$, which corresponds to an ellipticity equal to zero) includes the points that identify the states of polarization in which the circular component is null (linear polarization states). In general, the locus of points corresponding to the geographical parallels includes polarization states for which the ellipticity is constant ($2\chi = $ constant) while the locus of points corresponding to geographical meridians includes polarization states for which the orientation angle of the major axis of the polarization ellipse is constant ($2\psi = $ constant).

Since the sphere is normalized to S_0, if the degree of polarization of light is not unitary, its representation is a sphere with radius less than one: therefore, the Poincaré sphere also allows the description of light with a degree of polarization smaller than one.

The Poincaré sphere is a compact and powerful representation of polarization because it permits:

- to monitor graphically the evolution of the polarization state
- to bind the electromagnetic representation of the polarization state (the polarization ellipse) and photonics representation
- to represent effectively also not completely polarized light

2.2.3 Mueller Matrix

Mueller in 1943 [13, 14] described the same optical systems as Jones by using the Stokes calculus with a 4×4 matrix representation (composed by only real components), which can describe totally or partially polarized light. The Mueller matrix describes linear optical elements. An important feature of the Mueller matrix $\mathbf{M_M}$ for optical elements, with no Polarization-Dependent Loss (PDL), is that it is an orthogonal matrix, which has these important properties:

- determinant equal to one: $\det(\mathbf{M_M}) = 1$
- its inverse is equal to its transpose: $M_M^T = M_M^{-1}$

The Mueller matrix $\mathbf{M_M}$ for a polarization-altering device, such as the fiber, is defined as the matrix which transforms an incident Stokes vector $\mathbf{S}$ into the output Stokes vector $\mathbf{S'}$:

$$S' = \begin{bmatrix} S'_0 \\ S'_1 \\ S'_2 \\ S'_3 \end{bmatrix} = M_M S = \begin{bmatrix} m_{00} & m_{01} & m_{02} & m_{03} \\ m_{10} & m_{11} & m_{12} & m_{13} \\ m_{20} & m_{21} & m_{22} & m_{23} \\ m_{30} & m_{31} & m_{32} & m_{33} \end{bmatrix} \begin{bmatrix} S_0 \\ S_1 \\ S_2 \\ S_3 \end{bmatrix}. \tag{2.36}$$

$\mathbf{M}_M$ (k, λ) for an optical element is always a function of the direction of propagation k and wavelength λ. The Mueller matrix is an appropriate formalism for characterizing polarization measurements because it contains within its components all the polarization properties. When the Mueller matrix is known, then the output polarization state is known for an arbitrary incident polarization state. Further, for depolarizing optical systems, the Mueller calculus is superior to the Jones calculus because every Jones matrix can be expressed as a Mueller matrix but not always vice versa as in the case of a depolarizer [15].

2.2.4 Four-Dimensional Representation

The electromagnetic field propagating in a SSMF can be represented by a four-dimensional vector whose components are the phase and quadrature terms of the two polarization components of the field [16]:

$$E = E_x\,\hat{x} + E_y\,\hat{y} = \left[\left(x_1 + ix_2 \right)\hat{x} + \left(x_3 + ix_4 \right)\hat{y} \right]. \tag{2.37}$$

The Jones description (see Section 2.2.1) can be reconducted to an equivalent four-dimensional description, by expressing the complex Jones vector as the real four-dimensional vector:

$$E = \begin{pmatrix} \mathrm{Re}(E_x) \\ \mathrm{Im}(E_x) \\ \mathrm{Re}(E_y) \\ \mathrm{Im}(E_y) \end{pmatrix} = \begin{pmatrix} x_1 \\ x_2 \\ x_3 \\ x_4 \end{pmatrix}. \tag{2.38}$$

It is possible to derive the Stokes parameters, Equations (2.25)–(2.28), from the two complex components of the Jones vector, or, in the same way, from the four real components of the vector in Equation (2.38) [17]:

$$S_1 = E_x E_x^* - E_y E_y^* = x_1^2 + x_2^2 - x_3^2 - x_4^2 \tag{2.39}$$

$$S_2 = E_x E_y^* + E_x^* E_y = 2x_1 x_3 + 2x_2 x_4 \tag{2.40}$$

$$S_3 = i\left(E_x E_y^* - E_x^* E_y \right) = 2x_1 x_4 - 2x_2 x_3. \tag{2.41}$$

2.2.5 Evolution of Polarization in Optical Fiber

During its propagation along the optical fiber (z axis direction), the electric field E assumes a specific local polarization (SOP) at each point (x, y, z). The evolution of any initial SOP can be represented by a trajectory C(z) on the Poincaré sphere (only on the surface if there is no PDL). The local velocity and the direction of evolution, C′ = dC/dz, are governed by the coupled wave Equations (2.7)–(2.8). In order to find the evolution of polarization over a longer piece of fiber, these equations must be integrated. The representation on the Poincaré sphere is well suited for this integration, because, in the two-mode case, Equations (2.7) and (2.8) describe simply a rotation of the sphere with a certain angular velocity $\boldsymbol{\beta}(z)$, which represents the local birefringence vector in any point of optical fiber's propagation axis (z axis). $\boldsymbol{\beta}(z)$ can be expressed in terms of the coupled coefficients k_{ij} [18]:

$$\beta = |\beta| = \left[\left(k_{xx} - k_{yy} \right)^2 + 4k_{xy}k_{yx} \right]^{1/2} \tag{2.42}$$

$$2\psi = \tan^{-1}\left[\left(k_{xx} - k_{yy} \right) \middle/ \left(4k_{xy}k_{yx} \right)^{1/2} \right] \tag{2.43}$$

$$2\chi = \arg\left(k_{xy} \right). \tag{2.44}$$

Knowing $\boldsymbol{\beta}(z)$, for each dz, it would be possible to construct for any given input SOP the trajectory C(z) on the Poincaré sphere. This geometrical construction of C(z) as a succession of rotations permits a clear insight into the various polarization effects. The laws of rotation assume a particularly simple form when the rotations are infinitesimally small. In this case these laws allow simple geometrical interpretations in Stokes space (on the Poincaré sphere).

Consider the change of the polarization of light at fiber location z due to a small length addition dz of fiber. As described above, the change of the SOP, during its evolution along the fiber, is influenced by the fiber's local birefringence characterized by its effective relative dielectric tensor ε. This change is governed by the wave Equation (2.5) in which the term ε can be expressed in the following way [17]:

$$\varepsilon k_0^2 = \beta_0^2 I + \beta_0 \beta \sigma = \beta_0^2 I + \beta_0 \begin{pmatrix} \beta_1 & \beta_2 - i\beta_3 \\ \beta_2 + i\beta_3 & -\beta_1 \end{pmatrix}, \tag{2.45}$$

where β_0 is the common propagation constant, I the identity matrix and $\boldsymbol{\sigma} = (\sigma_1, \sigma_2, \sigma_3)$ the Pauli spin vector in Stokes space. The coefficients β_i of the expansion are the components of the local birefringence vector $\boldsymbol{\beta}(z)$ in Stokes space. According to the assumptions made in [17], it is possible to obtain the adiabatic wave equation for the Jones vector as expressed in (2.20):

$$\frac{d\begin{pmatrix} E_x \\ E_y \end{pmatrix}}{dz} + \frac{1}{2}i\beta\sigma\begin{pmatrix} E_x \\ E_y \end{pmatrix} = 0. \tag{2.46}$$

Finally, starting from the Equation (2.46), the law of infinitesimal rotation for birefringence is obtained as described below:

$$\frac{dS}{dz} = \beta \times S. \tag{2.47}$$

The law expressed in (2.47) is fundamental for the study of the evolution of polarization in optical fiber because it implies that an infinitesimal length dz of fiber rotates all SOPs about the direction $\boldsymbol{\beta}$ through an angle βdz. Thus Equation (2.47) describes the spatial evolution of the Stokes vector.

2.3 POLARIZATION EFFECTS IN LIGHTWAVE SYSTEMS

2.3.1 Polarization Mode Dispersion (PMD)

Polarization effects in optical fibers have been known for a long time. But their importance has been grown since technology improvements have:

1. enabled to realize long haul optical paths (with the introduction, e.g., of optical amplifiers). Obviously, the longer the path is the more optical elements may be present. The consequence is that also small effects like PMD (Polarization Mode Dispersion) and PDL (Polarization-Dependent Loss) become relevant;
2. approached throughput limit of optical fiber.

2.3.2 PMD Models

There are two fundamental models to simulate PMD in optical fibers: a high coherence model or a low-coherence model. The latter is usually referred to as the coupled-power model [19]. The most widely used high coherence model, which was developed specifically for dealing with PMD in single-mode fiber, is the Principal States Model [20].

The coupled-power model predicts that in the long length regime the net output pulse will be Gaussian in shape and broadened relative to the input pulse by an amount proportional to the square root of the waveguide length L and the Differential Group Velocity DGV of the two waveguide modes [19].

Figure 2.6 Pulse broadening caused by PMD.

The Principal States Model assumes that the optical loss in the fiber does not depend on the polarization and that the coherence time of the source is greater than the PMD-induced time shifts involved. For a digital lightwave system, this latter assumption is equivalent to assuming that the net time delay caused by PMD in the span is small compared with the bit period. The model thus addresses the regime in which most lightwave systems are expected to operate.

2.3.3 PMD and PDL Impairments

In spite of the optical fiber utilization advantages, it is necessary to consider also undesirable effects, which are present in real non-ideal optical fiber. In telecommunication and sensors application field, the presence of inherent and induced birefringence is crucial. The presence of birefringence may cause an undesirable state of polarization change. In the case of high-speed data transmission on long distances the PMD may occur. Due to this effect the light pulses are broadened. This may result in inter-symbol interference. While the inherent circular birefringence is negligible in common single-mode fibers, the inherent and induced linear birefringence may be present in considerable way. The inherent linear birefringence is mostly undesirable effect. Two important polarization effects in a fiber-optic communication system are PMD [21] and PDL (or Polarization-Dependent Gain [PDG]) [22].

PMD leads to broadening of optical pulses because of random variations in the birefringence of an optical fiber along its length. It causes the output SOP of a fully polarized input signal to vary with frequency. With the advent of long-distance high bit rate optical systems, PMD has become an important source of limitation for the system performance. In a first-order approximation, PMD, that is described by a Differential Group Delay (DGD) whose value is indicated with $\Delta\tau$ (see Figure 2.6) between two orthogonal Principal States of Polarization (PSPs), causes an undesired output pulse broadening; the frequency dependence of DGD and PSPs produces other distorting effects, considered as higher-order PMD effects [17].

PMD is difficult to compensate because of its randomness. The DGD, which represents a limit for the maximum system transmission rate, is expressed by:

$$\Delta\tau = L\frac{d}{d\omega}\left(\beta_x - \beta_y\right) = L\frac{\Delta n_{eff}}{c} + \frac{\omega}{c}\frac{d\Delta n_{eff}}{d\omega} \tag{2.48}$$

where L is the fiber's length and $\Delta n_{eff} = n_x - n_y$. By defining the dispersion vector $\boldsymbol{\Omega}$:

$$\Omega = \Delta\tau\, p \qquad (2.49)$$

where $\mathbf{p}$ is the unitary Stokes vector associated with one of the two PSPs of the fiber, it is possible to obtain the law of infinitesimal rotation:

$$\frac{dS}{d\omega} = \Omega \times S. \qquad (2.50)$$

The geometrical interpretation of this law is a rotation of the output Stokes vector on the Poincaré sphere as ω changes. The rotation axis is the PSP $\mathbf{p}$ and the rotation rate is the DGD $\Delta\tau$.

The dynamical PMD equation results from a combination of the infinitesimal rotation laws for PMD (2.50) and birefringence (2.47):

$$\frac{d\Omega}{dz} = \frac{d\beta}{d\omega} + \beta \times \Omega. \qquad (2.51)$$

Equation (2.51) is a general representation of the evolution of the dispersion vector in SSMF and is based solely on the assumption of negligible PDL [23]. PMD limits the system performance when the single-channel bit rate is extended to beyond 10 Gbit/s.

Several techniques have been developed for PMD compensation in dispersion-managed lightwave systems; they can be classified as being optical or electrical. Figure 2.7 shows the basic idea behind the electrical and optical PMD compensation schemes [24].

PDL and PDG usually occur if a lightwave system contains multiple elements that amplify or attenuate the two polarization components of a pulse differently; in this way the polarization state is easily altered.

PDL in optical systems has the potential of considerably degrading system performance by reducing the Signal-to-Noise Ratio (SNR) [25, 26].

The noise enhancement is generated by two mechanisms:

- One is related to the noise emitted directly into the polarization state of the amplified signal.
- The other is related to the noise emitted into an orthogonal polarization state.

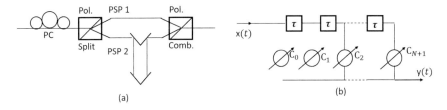

Figure 2.7 Schemes of (a) optical and (b) electrical PMD compensators.

The Optical SNR (OSNR) penalty caused by the first mechanism can be mitigated by including periodic channel-by-channel gain equalization along the link. In principle, its effect can be completely eliminated if such equalization is implemented at every amplifier location. The second mechanism, related to the orthogonal noise, is critical because in the presence of PDL some of the orthogonal noise is coupled back into the polarization of the signal in the process of propagation. This portion of the orthogonal noise beats with the signal as it impinges upon the photodetector, giving rise to a potentially significant degradation in performance. The deterioration in performance due to the noise effect of PDL can be quantified in terms of the deterioration in the electric SNR, which is equivalent in high OSNR systems to the deterioration in the ratio between the optical signal power and the power of the noise that is parallel to it at photodetection [22].

2.3.4 Birefringence

As mentioned in the previous paragraph, an ideal single-mode fiber with circular core cross-section along its length, made from homogenous isotropic material, will exhibit the same refractive index $\bar{n}$ for both of the modes. At this condition, the modes remain degenerate. Birefringence is introduced in a fiber whenever the circular symmetry of the ideal fiber is broken, thus producing an anisotropic refractive-index distribution in the core region.

The fiber will behave as a birefringent medium with different refractive indices n_x and n_y. Birefringence is therefore an optical fibers' natural feature. It can result from either a geometrical deformation of the core or a material anisotropy through various elasto-optic, magneto-optic or electro-optic index changes. This implies that the birefringence may be generated by production defects and/or external perturbations. The majority of these causes introduce a linear birefringence while the circular birefringence has a smaller number of causes. In order to understand birefringence's meaning, it is important to distinguish between linear and circular birefringence.

Linear birefringence has several origins: bendings, geometrical imperfections, and stress-induced anisotropies. All of them act randomly along optical fibers.

Circular birefringence can be caused instead by an external magnetic field aligned with the axis of propagation, or by twisting the fiber itself.

2.3.5 Linear Birefringence

The linear birefringence may be of latent or induced nature. Linear birefringence $\Delta\beta_l$, expressed in rad/m, is:

$$\Delta\beta_l = 2\pi\left|n_x - n_y\right|/\lambda. \tag{2.52}$$

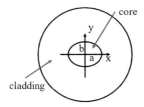

Figure 2.8 Elliptical cross-section of the non-ideal fiber core.

Linear birefringence caused by production defects:

1. The main cause of linear birefringence in real fiber is the manufacture imperfection. The first source is the geometrical anisotropy of the cross-section of the fiber core that is not ideally circular but slightly elliptical, as shown in Figure 2.8.

   Generally, this birefringence cannot be represented by a simple expression as it depends strongly on the frequency, or V-value, at which the fiber is being operated. However, if only frequencies near the higher-mode cut-off are considered (V ≈ 2.4 for step-index fibers), the birefringence for fibers with small-core ellipticity, (a/b − 1) ≪ 1, can be approximated by [9]:

$$\beta_c \approx 0.2\,k_0\left(\frac{a}{b}-1\right)(\Delta n)^2, \tag{2.53}$$

   where $\Delta n = n_1 - n_2$ is the refractive index difference between the core and cladding regions [27]. The fiber core ellipticity is not the single source of fiber birefringence imposed by the manufacture.

2. A second important source, which may take effect, is the presence of asymmetrical transverse stress. It introduces a linear birefringence via elasto-optic index changes. The stress may be frozen internally in the fiber during fabrication as the result of different thermal contraction between differently doped non-circularly symmetric regions of the fiber. The usually more heavily doped non-circular cladding is constrained by the outer (often silica) fiber jacket or substrate tube to produce the stress asymmetry. If this cladding region is elliptical with major and minor diameters of 2A and 2B respectively and the core is perfectly circular, this birefringence is [28, 29]:

$$\beta_S = \frac{C_S}{\left(1-\upsilon_p\right)}\Delta\alpha\Delta T\frac{A-B}{A+B}, \tag{2.54}$$

where:

$$C_S = 0.5 k_0 n_0^3 (p_{11} - p_{12})(1 + \upsilon_P).$$

(2.55)

The ratio C_S/k_0 is often loosely referred to as the strain-optical coefficient. The parameter n_0 is the mean refractive index of the fiber, p_{11} and p_{12} are the components of the strain-optical tensor of the fiber material and υ_p is the Poisson's ratio. The term $\Delta\alpha = \alpha_{cl} - \alpha_{st}$ is the difference between the expansion coefficients of the cladding α_{cl}, and the substrate α_{st} materials and $\Delta T = T_r - T_s$ is the difference between the room temperature T_r and the softening temperature T_s of the more heavily doped material. This stress birefringence is constant everywhere within the elliptical cladding and, provided $B \gg \rho$ (with ρ the core radius), the mode field is confined within the elliptical cladding. Thus, this birefringence is independent of the V-value.

Consider the case where, due to an imperfect technology process of fiber drawing from hot preform, elliptical density distribution is present. The far area of the cladding influences the inner area by centripetal pressure after the fiber cooled down. Since the core-close area has a non-homogenous density, the pressures on core, p_x and p_y will act non-uniformly as illustrated in Figure 2.9.

Due to the photo-elastic effect, which causes pressure dependent anisotropy, the fiber core becomes a birefringent medium. In common single-mode fibers, the influence of inner stress induced linear birefringence is weak in comparison with the birefringence owing to elliptical core cross-section.

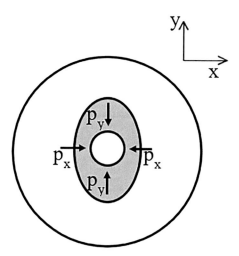

Figure 2.9 Non-uniform stress on fiber core owing to imperfect inner structure.

The same stress birefringence exists in a fiber with an elliptical core, with 2a and 2b replacing 2A and 2B. However, now, even though the stress asymmetry is constant within the elliptical core, it varies rapidly outside the core. In this case, the mode field is not contained within the ellipse; thus, this stress birefringence depends upon the V-value and is given by [29, 30]:

$$\beta_{SC} = \left(1 - \frac{u^2}{V^2}\right)\frac{C_S}{\left(1 - \upsilon_p\right)}\Delta\alpha\Delta T\frac{a - b}{a + b}, \qquad (2.56)$$

where $V = k_0 b\left(n_1^2 - n_2^2\right)^{1/2}$ and $u = k_0 b\left(n_1^2 - n_b^2\right)$ are the normalized parameters, n_b is the effective mode index along the minor diameter, $\Delta\alpha = \alpha_{co} - \alpha_{cl}$ and α_{co} is the expansion coefficient of the core.

Linear birefringence caused by external influences:

Linear birefringence in SSMF may also be induced by outer influences. It is caused by outer mechanical stress (pressure or tensile force) on fiber cladding or by applying an external electric field. Cladding transfers the mechanical stress on the core and a similar effect to that described above happens. In practice, an action of force in one dominant direction appears usually. It induces two different refractive indices n_x and n_y for the axes of symmetry, x and y.

1. One of the possible external causes that can generate linear birefringence is fiber bending, which is illustrated in Figure 2.10.
   The same type of stress birefringence is induced in a fiber when it is freely bent, bent under tension around a drum or kinked over a sharp object on the drum.

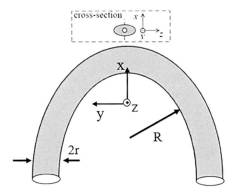

Figure 2.10 Geometric relation of fiber bending causing induced linear birefringence.

Freely bending the fiber places the outer portion of the fiber cross-section in tension, which, then, presses laterally on the inner portion, which is in compression. The birefringence from this second-order transverse stress is [31]:

$$\beta_b = 0.5 C_S \frac{r^2}{R^2}, \qquad\qquad (2.57)$$

where R is the radius of the bend and r is the outer radius of the fiber.

Winding the fiber with axial tension F onto a drum introduces an additional linear birefringence, which adds to the bending birefringence. This tension-coiled birefringence β_{tc} results from the lateral force exerted by the drum on the fiber in reaction to the tensile force F. The expression of β_{tc} is [32]:

$$\beta_{tc} = C_S \frac{2 - 3\upsilon_P}{1 - \upsilon_P} \frac{r}{R} \varepsilon, \qquad\qquad (2.58)$$

where $\varepsilon = F/(\pi r^2 E)$ is the mean axial strain in the fiber, with E that represents the Young's modulus of the fiber material.

If the drum has a sharp ridge of height H protruding from it when the fiber is wound with tension F, then the transverse stresses are redistributed over the short fiber length in the vicinity of the ridge and an additional linear birefringence, which adds to both β_b and β_{tc}, is introduced [33]. This kink birefringence β_k results from both the increased curvature of the fiber near the kink and the localized lateral force F_0 exerted by the ridge on the fiber. An average value of β_k is:

$$\overline{\beta_k} = 0.5 C_S \frac{r}{R} \sqrt{\frac{H\varepsilon}{R}}. \qquad\qquad (2.59)$$

The practical meaning of this kink birefringence is that it will be introduced whenever one fiber turn crosses another in a fiber coil.

2. A second significant outer cause, which induces linear birefringence in the fiber, is a lateral pressure, as shown in Figure 2.11. Induced anisotropy in the fiber is a result of photo-elastic effect, which is generated by compressing fiber between parallel plates or into an angled V-groove. For two parallel plates, this birefringence is [34]:

$$\beta_F = 4 C_S \frac{F}{\pi r E}, \qquad\qquad (2.60)$$

where F is the linear force (force per unit length). This relationship holds for sharp (concentrated at a point) line forces. If the forces are

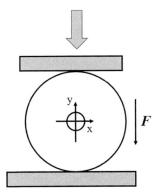

Figure 2.11 Imposing a lateral pressure force on the fiber.

distributed over part of the fiber surface, a reduction of up to a factor of $8/\pi$ can result.

When a fiber is pressed into a V-groove with an included angle of 2δ, the linear birefringence is [35]:

$$\beta_V = 2C_S\left(1 - \cos 2\delta \sin 2\delta\right)\frac{F}{\pi r E}. \tag{2.61}$$

The meaning of (2.61) is that it is not possible to clamp a fiber in a V-groove without introducing birefringence.

3. Finally, there should be mentioned another way to induce the fiber linear birefringence. It may be imposed by electro-optical effect in the fiber core. However, the fiber core is made from amorphous material and the electro-optic effect is very weak. A transverse electric field E_K, via the electro-optic Kerr effect, will introduce a linear birefringence in the fiber. In this case the birefringence is given by [18]:

$$\beta_E = 2\pi B_K E_K^2, \tag{2.62}$$

where B_K is the Kerr electro-optic constant of the fiber. For silica fibers, B_K is too small for practical significance other than for birefringence measurements via polarization modulation techniques.

2.3.6 Circular Birefringence

The presence of circular birefringence in the fiber results in polarization plane rotation. When the fiber is free from linear birefringence and a linear polarized wave is injected into the fiber itself, at the output there will be a linear polarized wave with a rotated polarization plane. The angle of plane rotation is due to the circular birefringence rate and to the fiber length.

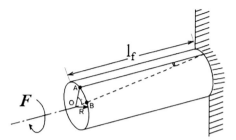

Figure 2.12 A fiber section with length l_f subjected to a torsion with angle τ.

The latent origin of circular birefringence in SSMFs is negligible with respect to that of linear birefringence. Nevertheless, it is possible to impose it in manufacturing process or to induce it by outer influence. A suitable mechanical stress or a magnetic field applied in the direction of fiber axis can be able to realize this kind of birefringence.

If a fiber section with length l_f is subjected to a twist with a specific rate τ':

$$\tau' = \frac{\tau}{l_f}, \tag{2.63}$$

where τ is the twist angle as shown in Figure 2.12, a sheer stress is imposed in the plane perpendicular to fiber axis.

Imposed sheer stress results in fiber core anisotropy owing to photo-elastic effect. In order to describe optical properties of anisotropic fiber core, it is useful to exploit tensor matrix of dielectric constant ε [36]:

$$\varepsilon = \varepsilon_i + \Delta\varepsilon_t = \begin{bmatrix} \varepsilon & -g\tau'y & 0 \\ g\tau'y & \varepsilon & -g\tau'x \\ 0 & g\tau'x & \varepsilon \end{bmatrix}, \tag{2.64}$$

where ε_i is the dielectric constant tensor of original medium and $\Delta\varepsilon_t$ is the tensor of twist contribution. Typical values of the stress-optic coefficient g are between 0.14 and 0.16, depending on the core dopants. In this case, circular birefringence $\Delta\beta_c$, expressed in rad/m, is:

$$\Delta\beta_c = g\tau'. \tag{2.65}$$

In a twisted fiber, the total birefringence $\Delta\beta$, expressed in rad/m, is:

$$\Delta\beta = \sqrt{\Delta\beta_l^2 + \left(g\tau' - 2\tau'\right)^2}. \tag{2.66}$$

The birefringence $\Delta\beta$ includes a linear part ($\Delta\beta_l$) and a circular part. In the circular part, $g\tau'$ is the physically induced value of the circular birefringence whereas $2\tau'$ accounts for the rotation of the local axes and is comparable to a "geometrical value" of the circular birefringence. Generally, in presence of both types of birefringence, the fiber behaves as phase retarder and polarization rotator simultaneously.

The second source of fiber circular birefringence is the magneto-optical effect. Between three types of magneto-optical effect (Cotton-Mouton, Kerr, Faraday) [37], the Faraday effect is significant for silica fiber. It induces circular birefringence owing to magnetic field action in direction along the fiber axis. Analogous to fiber twist, the Faraday magneto-optical effect modifies the dielectric constant tensor in the following way:

$$\varepsilon = \varepsilon_i + \Delta\varepsilon_{mo} = \begin{bmatrix} \varepsilon & -j\eta B & 0 \\ j\eta B & \varepsilon & 0 \\ 0 & 0 & \varepsilon \end{bmatrix}, \tag{2.67}$$

where $\Delta\varepsilon_{mo}$ is the tensor of magneto-optical effect contribution, B is the magnitude of flux density of the external magnetic field, η is a coefficient proportional to the magneto-optic specific rotation coefficient [38].

In order to explain the origin of Faraday magneto-optical effect, it is possible to model the effect as an electron oscillator movement subjected to a magnetic field [39]. The effect itself results from interaction of outer magnetic field with oscillating electron, which is excited by the electric field of the lightwave. Electrons represent harmonic oscillators. For them equations of forced oscillations hold.

As in the case of circular birefringence induced by twist, it is possible to derive an expression for circular birefringence induced by an external magnetic field; in particular, the following relationship holds:

$$\Delta\beta_c = V B, \tag{2.68}$$

where V is the Verdet constant, which characterizes the magneto-optic properties of the medium. It depends on the wavelength. The right part of Equation (2.68) is the basic relation for Faraday magneto-optic effect. The effect is non-reciprocal. The polarization rotation direction depends on the mutual orientation of magnetic flux density B and the wave propagation direction. The polarization of wave propagating in the direction of B experiences a rotation of $\Delta\beta_c\, l_f$. The polarization of wave propagating in the opposite direction to B experiences a rotation $-\Delta\beta_c\, l_f$.

2.3.7 Birefringence Models

In the past, various models were proposed in order to simulate birefringence effects on electromagnetic field propagation along optical fibers. Many of these works focused on the study of linear birefringence spatial evolution, considering that in standard telecommunication optical fibers, it is generally supposed that the birefringence can be treated as purely linear and circular birefringence may be considered negligible. In particular, there are three fundamental theoretical models able to describe the birefringence behavior [40]. They all refer to the properties of the local birefringence vector $\boldsymbol{\beta}$ (β_1, β_2, β_3). The terms β_1 and β_2 represent linear birefringence while β_3 takes into account of circular birefringence.

These models can be summarized as follows:

1. The birefringence is assumed to comprise two terms; one is constant and deterministic and the other is a white-noise process that describes the random perturbations [23].
2. It is assumed that the modulus of the birefringence is fixed, whereas its orientation varies according to a Wiener process [41].
3. The birefringence vector varies both in modulus and in orientation. The first two components of the local birefringence vector are assumed to be Gaussian random processes, statistically independent of each other, with zero mean and the same standard deviation [41].

In the first model it is $\beta_3 << \beta$. Hence β_3 is a negligible term, whereas for models 2 and 3 it is assumed that the fiber is not affected by circular birefringence ($\beta_3 = 0$).

Among all the models of birefringence known in the literature, for SSMFs with linear birefringence, only the third model proposed by Wai and Menyuk in [41], described below (see paragraph 2.3.8), is consistent with the experimental evidence [40]. The most suitable model to describe SSMFs with twist-induced circular birefringence is described in paragraph 2.3.9.

2.3.8 Model for SSMF with Linear Birefringence

Physically, PMD has its origin in the birefringence that is, as described in the previous paragraphs, a natural characteristic of the optical fiber. While this birefringence is small in absolute terms in communication fibers, with values of $\Delta\beta \approx 10^{-7}$, the corresponding beat length L_B is only about 10 m (see paragraph 2.2) – far smaller than the dispersive or nonlinear scale lengths which are typically hundreds of kilometers or more – so that the birefringence should be considered large. This large birefringence could be devastating in communication systems; nevertheless, the orientation and the strength of the birefringence are randomly varying on a length scale that is on the order of 100 m. The rapid variation of the birefringence tends to make the effect of its average close to zero. The residual effect leads to pulse spreading, referred to PMD. The randomly variations of the orientation and the strength of the birefringence cause scattering light from one local polarization eigenstate to another.

The key physical parameters that determine the rate at which light pulses spread in linear systems [42] are the polarization decorrelation length h_E, the average beat length L_B and the autocorrelation length of the birefringence fluctuations in the fiber, h_{fiber} [43].

The parameter h_E, which depends on both L_B and h_{fiber}, is the length scale over which the electric field loses memory of its initial distribution between the local polarization eigenstates and can be treated as random. The fiber autocorrelation h_{fiber} is the length over which an ensemble of fibers, all of which initially have the same orientation of the axes of birefringence, loses memory of this initial orientation.

The model described in [41] allows both the birefringence strength and orientation to vary in accordance with a bi-Maxwellian distribution. According to this model, the linear birefringence components β_1 and β_2 are independent Langevin processes (i = 1, 2):

$$\frac{d\beta_i}{dz} = -\rho\beta_i(z) + \sigma\eta_i(z),$$ (2.69)

with $\eta_1(z)$ and $\eta_2(z)$ that represent independent white noise processes with the following statistic properties (i = 1, 2):

$$E\left[\eta_i(z)\right] = 0,$$
$$E\left[\eta_i(z)\eta_i(z+u)\right] = \delta(u),$$ (2.70)

$\delta(u)$ being the Dirac distribution. Linear birefringence is a stationary stochastic process [44]. The terms ρ and σ in Equation (2.69) represent statistical properties of the birefringence (the key physical parameters described above).

In particular, from [41] and [45], it is possible to derive the following relationships:

$$\rho = \frac{1}{h_{fiber}}, \sigma = \frac{4}{L_B}\sqrt{\frac{\pi}{h_{fiber}}}.$$ (2.71)

2.3.9 Model for Twisted SSMF

As it has been described in paragraph 2.3.6, in particular cases, circular birefringence shows its effects in optical fibers. In these contexts, the term β_3 of the local birefringence vector β can no longer be neglected. When a general birefringent fiber is twisted (Figure 2.13), the trajectory $C(z)$ is complicated by the fact that $\boldsymbol{\beta}(z)$ is not constant but is itself moving on the Poincaré sphere.

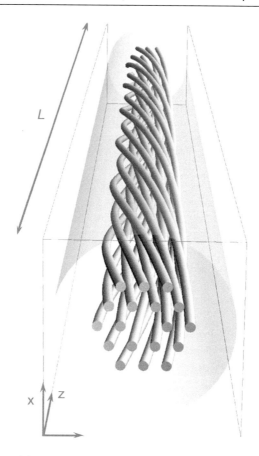

Figure 2.13 Twisted fiber.

The model of [45], described in this paragraph, shows the birefringence behavior when the optical fiber is subjected to a twisting process. With this model, the local birefringence vector can be expressed in the following way:

$$\beta(z) = T(z)\begin{pmatrix} \beta_1 \\ \beta_2 \\ \beta_3 \end{pmatrix} = T(z)\begin{pmatrix} \beta_1 \\ \beta_2 \\ g\tau'(z) \end{pmatrix}. \tag{2.72}$$

$T(z)$ is the rotation matrix of the cross-sectional plane z originated by the twisting process:

$$T(z) = \begin{pmatrix} \cos 2\tau(z) & -\sin 2\tau(z) & 0 \\ \sin 2\tau(z) & \cos 2\tau(z) & 0 \\ 0 & 0 & 1 \end{pmatrix}, \tag{2.73}$$

while $\tau(z)$ is the twist measured in radians and $\tau'(z)$ is the twist rate expressed in rad/m (see paragraph 2.3.6). The parameter g refers to optical fiber coupling parameters and represents the proportionality coefficient between twist rate and induced circular birefringence, that is $\beta_3(z) = g\tau'(z) = -2ik_{xy}$. In fact, circular birefringence is characterized by an imaginary k_{xy}:

$$k_{xy} = -k_{yx} = -in_0^2 p_{44}\tau'/2 = ig\tau'/2 \qquad (2.74)$$

where n_0 is a mean refractive index of the fiber and p_{44} is a component of the elasto-optic tensor p_{rs} and $g = -n_0^2 p_{44}$ [18]. Equation (2.72), together with (2.69), defines the birefringence model of a twisted fiber.

When the fiber is twisted, $\beta(z)$ moves along a parallel circle on the Poincaré sphere, and the trajectories become cycloidal curves [18]. These trajectories can be ordinary cycloids, curtate cycloids or prolate cycloids depending on the twist rate (Figure 2.14). To understand these trajectories, it is possible to note that the motion of $\beta(z)$ can be expressed by the vector product:

$$\beta'(z) = 2\tau' \times \beta(z). \qquad (2.75)$$

The twist vector is directed parallel to the polar axis. The analysis of a twisted fiber is much simplified if it is performed with respect to a moving frame, rotating at twice the twist rate.

Nevertheless, it is convenient to make some more assumptions on twist properties; in particular, it is fundamental to know whether twist's variations are random or deterministic. It is hard to determine how external perturbations might induce a randomly varying twist along the fiber. On the contrary, twist might be deterministically induced while fiber is either wound on or unwound from a bobbin. In the scenario of a deterministic twist, it is necessary to find the function that can better approximate the behavior of the twist itself.

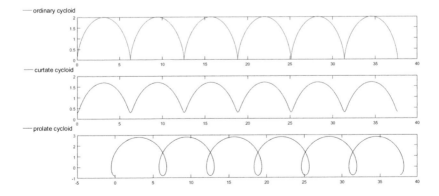

Figure 2.14 Types of cycloids.

2.3.10 SOP Fluctuations Compensation

Because the received field SOP fluctuates, due to the fiber's birefringence, the values of S_1, S_2 and S_3 of each transmitted symbol change in time and must be tracked by the receiver. This can be done by exploiting the property of the SSMF that the output SOP evolution can be represented as a rotation of the Poincaré sphere. In this paragraph it will be described the solution for SOP fluctuations compensation proposed in [10].

At the beginning of the transmission, a synchronization pattern, composed of N, replicas of each symbol, allows the reference SOPs on the receiver Poincaré sphere to be identified.

The tracking algorithm used during the transmission is based on the property that the characteristic time T_{SOP} of the SOP fluctuations due to fiber birefringence is on the order of minutes [46]. In a point-to-point link this effect constitutes the only cause of SOP fluctuations. On the other hand, in a multichannel network, switching among different channels introduces further fast SOP changes. In this case, a system reset procedure is needed which must be foreseen in the network communication protocol.

In the presence of slow fluctuations, the reference SOPs, which are associated with the possible transmitted symbols, must be updated every T_{UP} seconds so that $1/W \ll T_{UP} \ll T_{SOP}$, where W is the signal bandwidth.

The polarization tracking circuit uses N memory cells, each one of which records the three Stokes parameters of a reference SOP. Between updates and for each time slot, the decision device estimates the received symbol by associating it with one of the possible transmitted symbols. The received Stokes parameters are then added to the memory cell containing the Stokes parameters of the reference SOP corresponding to the received symbol. This operation averages the values of the received Stokes parameters corresponding to each transmitted symbol. At the end of the updating interval each reference SOP is replaced by the mean Stokes parameters recorded in the appropriate memory cells.

By expressing the estimated Stokes parameters of the ith reference SOP $(k = 1, 2, 3)$ as:

$$S_{k,i} = S_{k,i}^a + n_{k,i} \tag{2.76}$$

where $S_{k,i}^a$ represent their actual values and $n_{k,i}$ are independent Gaussian error terms whose variance is given by $2/T_{UP}$, the expression of the decision variables can be written as:

$$q_i = \sum_{k=1}^{3} S_k S_{k,i}^a + \sum_{k=1}^{3} S_k n_{k,i} + \sum_{k=1}^{3} N_k S_{k,i}^a + \sum_{k=1}^{3} N_k n_{k,i}. \tag{2.77}$$

Since $1/W \ll T_{UP}$, it follows that:

$$\sum_{k=1}^{3} S_k^2 n_{k,i}^2 \ll \sum_{k=1}^{3} N_k^2 S_{k,i}^{a2}$$

$$\sum_{k=1}^{3} N_k^2 n_{k,i}^2 \ll \sum_{k=1}^{3} N_k^2 S_{k,i}^{a2}$$

(2.78)

and:

$$q_i = \sum_{k=1}^{3} \left(S_k + N_k \right) S_{k,i}^a$$

(2.79)

so that the reference SOP estimation can be considered ideal and only the noise terms N_k determine the system performance.

A simple implementation of the above tracking algorithm is shown Figure 2.15. It is based on three switches and 3N integrators. After the

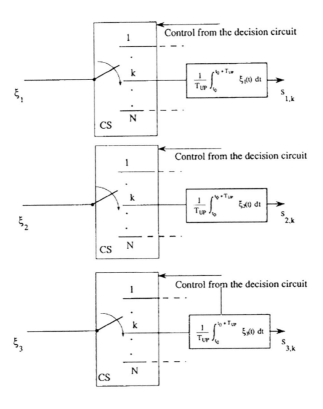

Figure 2.15 Reference SOP estimation circuit (CS = controlled switch).

received symbol has been identified, the switches are set to charge the integrators of the corresponding reference SOP so that at the update time the output voltage of all the integrators can be used to calculate the mean reference Stokes parameters. After the update, the capacitors are discharged for a new estimation for the reference SOPs.

2.4 NOVEL MULTILEVEL PoLSK COHERENT SYSTEM

This paragraph illustrates, in detail, a novel Multilevel PolSK scheme [47]. The theoretical concepts behind this model, well suitable for LANs, are described. It is explained the adopted simulation model, useful for the study of the behavior of the innovative modulation scheme. Then, a detailed statistical analysis is illustrated. Moreover, a possible reference mathematical model is proposed. It is also given a description of the initial experimental steps. Finally, the system performance is analyzed in the presence of the main techniques of detection (coherent and direct).

2.4.1 Birefringence as a Resource

The PolSK modulations are affected by the birefringence, a peculiar optical fiber characteristic, as thoroughly illustrated above. The birefringence affects deeply the evolution of the SOP of the optical signal along the fiber. In fact, birefringence causes random SOP rotations during signal propagation along optical fiber; therefore, when adopting a polarization modulation, it is necessary to implement, at the receiver end, birefringence acquisition and tracking systems to recover exactly the transmitted SOP (see paragraph 2.3.10).

Birefringence has always been seen as a disadvantage for two main reasons:

- It originates the PMD, which represents a severe constraint for high bit rate systems.
- It generates SOP fluctuations, which are harmful for PolSK modulation.

However, as in many real cases, the same phenomenon, always considered as a drawback, can be exploited advantageously in other contexts.

We want to describe a new concept of multilevel polarization modulation, which is able to exploit optical fiber birefringence as an advantage over propagation distances comparable to those used in LANs. It will be shown that, at the origin of the proposed model, there is the exploitation of induced circular birefringence by means of twisting process of an optical fiber.

The original idea has been to follow the approach of fiber twisting (described in the paragraph 2.3.6), which may impose a strong circular birefringence in the core. If the rate of induced circular birefringence was much greater than the rate of linear birefringence, relation (2.66) could be modified in such a way that the influence of circular birefringence could be able to dominate the linear birefringence.

Twisted fiber will behave as polarization plane rotator, which preserves the polarization state of the wave [48]. The circular birefringence in fiber core is possible by imposing a fiber twisting process in a plane, which is perpendicular to fiber longitudinal axis. The rate of circular birefringence corresponds to photo-elastic properties of the core, core refractive index, fiber length and specific twist rate as illustrated by the relation (2.63). In order to minimize the influence of linear birefringence, the specific twist rate should be maximized. However, this process is limited by torsion limit of the fiber. When exceeds, the fiber may be broken.

The fabrication process of twist fiber is therefore very sensitive; an approximate torsion limit of common SSMF was calculated in 100 turns per 1 meter by Payne et al. [27].

2.4.2 System Description

The Stokes formalism for the polarization representation (see paragraph 2.2.2) is adopted for the analysis and the simulation.

Starting from the simulative results relative to the birefringence model described in the paragraph 2.3.9, it has been possible to define a system configuration for high capacity LAN.

First of all, the spatial evolution of the SOP of an electromagnetic field, which travels along a twisted fiber, has been analyzed according to the Stokes law of motion (2.47) with the local birefringence vector given by (2.72). In particular, simulations have been obtained by dividing Equation (2.47) into three different differential equations, one for each component of the Stokes vector:

$$\frac{dS_1}{dz} = -\beta_3 S_2 + \beta_2 S_3 \tag{2.80}$$

$$\frac{dS_2}{dz} = \beta_3 S_1 - \beta_1 S_3 \tag{2.81}$$

$$\frac{dS_3}{dz} = -\beta_2 S_1 + \beta_1 S_2, \tag{2.82}$$

where β_1, β_2 and β_3 are the components of the local birefringence vector given by (2.72). In order to solve in the space domain the three differential Equations (2.80)–(2.82), it has been chosen the *MATLAB®* Ordinary Differential Equation (ODE) Solver [49]. ODE contains one or more derivatives of a dependent variable with respect to a single independent variable. There exist various different ODE Solvers in *MATLAB*. The adopted choice of this book was *ode45* that is the most common solver because it performs well with most ODE problems. Moreover, after several attempts, it has proven to be the most suitable for the adopted model.

A standard waveplate model has been adopted with a waveplate length of 0.01 m. Numerical simulations have been repeated 1000 times, by adopting a serial calculation method, for building up a suitable statistical ensemble. It has been assumed a fully polarized lightwave which propagates along a fiber with no PDL. In addition, the total power field density S_0 is normalized to 1 (radius of the Poincaré sphere equal to 1) and all the signal SOPs lie on the surface of the Poincaré sphere.

A confirmation of the accuracy of this model is given by the fact that the spatial trajectories of the SOP are generally cycloids as established in [18] and [50] (see Figure 2.14). In the test simulation (Figure 2.16), in order to enhance the visualization of the cycloidal patterns, they have been chosen two weak values of the twist rate (2 rad/m for the black prolate cycloid and 4 rad/m for the red curtate cycloid). To model the twist in the test simulation, it has been adopted a linear twist (as a function of z) and, consequently, a constant twist rate (in this case induced circular birefringence is no longer a function of z) as described in [18] and [45].

Figure 2.16 shows two different trajectories assumed by a SOP whose initial state is the vertical polarization [−1,0,0]. It can be noted that the trajectories can be ordinary cycloids, curtate cycloids or prolate cycloids, depending on the twist rate (see also Figure 2.14). After testing the simulation approach, the attention has been focused on the build of a model that was able to simulate a local area network. The simulations have been extended to a distance of 2 km, without any optical joint, with the same step size of 0.01 m. Various values have been assumed for h_{fiber} and L_B as described in [45] without noting significant changes.

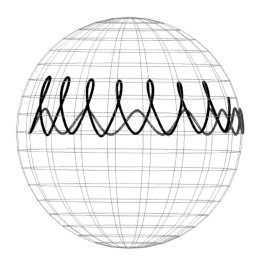

Figure 2.16 Trajectories of the SOP [−1,0,0].

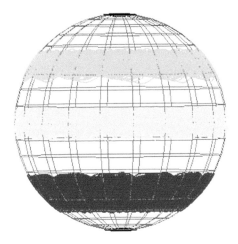

Figure 2.17 Spatial evolutions of the five symbols of the constellation.

Equations (2.47), (2.69) and (2.72) have been combined to form the complete Stochastic Differential Equation (SDE). Linear birefringence was numerically generated by integrating (2.22) for different random evolutions of $\eta_1(z)$ and $\eta_2(z)$, while induced circular birefringence was generated by testing different functions for the simulation of the twist evolutions along the propagation z axis: linear, sawtooth and sinusoidal. The function that has given the best results is the linear one, on which the simulative choice fell.

A constellation of five symbols (a symbol is identified with a SOP represented on the surface of the Poincaré sphere) has been simulated with a constant twist rate. Spatial evolutions of these symbols are shown in Figure 2.17.

It is worth noting how all the constellation symbols evolve within a specific "band of polarization" (physical region). The simulated multilevel modulation scheme provides five different parallel bands.

Only the component S_3 of the transmitted SOPs must assume a specific value to remain confined during spatial propagation within its own "band of polarization" (see Table 2.2). Once fixed S_3, the other two components

Table 2.2 Band latitude range of constellation points

Transmitted S_3	Upper boundary (degrees)	Lower boundary (degrees)	Bands latitude range (degrees)
1	90	80.9	9.11
0.5	41.91	21.67	20.25
0	7.28	−9.03	16.31
−0.5	−28.82	−55.83	27.01
−1	−80.9	−90	9.11

S_1 and S_2 of the Stokes vector can be set in transmission at any initial value. This important result allows to considerably simplify the transmitter.

Therefore, when the signal propagates along the z fiber axis, S_1 and S_2 can vary from -1 to 1, that is the whole range of possible values. On the contrary S_3 can assume only the values within its band latitude boundaries. It has been verified by simulation that the probability that S_3 exceeds these latitude boundaries is negligible. So, performance (see paragraph 2.4.7) of the received signal perturbed by Additive White Gaussian Noise (AWGN) is very similar to those obtained in [51]. The best performance, in terms of spatial confinement of the bands, was obtained, as mentioned above, by adopting a linear twist and, consequently, a constant twist rate.

In particular, with this configuration the fiber has been assumed to be twisted at a constant rate of 6.8 rad/m (about 1 turn/m). It is thus possible, by using a twisted fiber, to obtain a new concept of multilevel polarization modulation realized with a "fluid" constellation of parallel bands. By forcing the twisting process, the number of bands can be increased (bands boundaries become narrower) and consequently a greater single-channel throughput may be achieved with the same available bandwidth.

Therefore, an intrinsic feature of the optical fiber as the birefringence, with an appropriate fabrication process and for distances comparable with LAN, gives rise to a strict physical confinement of the SOP of the electromagnetic field within a "band of polarization" with established latitude boundaries. These latitude boundaries remain unchanged regardless of the behavior of linear birefringence and the fundamental parameter to be analyzed becomes the term S_3 of the SOP. As it can be seen from (2.31) to (2.32), this term is uniquely related to the latitude of the SOP. This implies that, at the receiver end, it is possible to know the transmitted SOP by measuring only the component S_3 of the received SOP. Depending on the "band of polarization" in which S_3 is located, it will be possible to identify the transmitted SOP because latitude boundaries are known. The free permitted variations of S_1 and S_2 make "fluid" this type of constellation that is no longer anchored to a rigid geometric structure.

The advantage of a "fluid" constellation lies in the fact that the decision regions may be associated directly with physical regions ("bands of polarizations"). The structure of the constellation is such that the modulator must only change the value of the S_3 component. The drawback is the presence of a limited number of polarization bands (physical tracks), conditioned by the twisting process that it is possible to introduce in the optical fiber.

This configuration requires, therefore, a Stokes receiver (Figure 2.18) much less complex than the conventional one, in which:

- It is necessary to reveal only the component S_3 of the received SOP.
- There is no need to implement a mechanism for tracking the birefringence.

Simplified Stokes Receiver

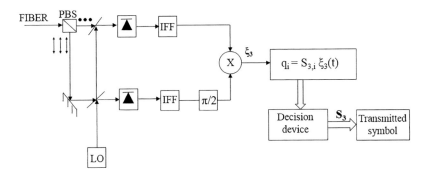

Figure 2.18 Simplified Stokes Receiver for "bands" polarization modulation (LO = optical local oscillator, IFF = intermediate frequency filter).

2.4.3 Statistical Analysis

This novel type of multilevel-polarization modulation ("bands of polarization" modulation) goes beyond the classical concept of symbols belonging to a rigid constellation in the Euclidean space (see also [47]). This modulation scheme takes advantage of intrinsic characteristics of optical fibers such as the birefringence. As a matter of fact, with a suitable twisting process, the induced circular birefringence β_3 becomes predominant with respect to the linear birefringence components β_1 and β_2. In this case, the evolution of the SOP during its spatial propagation along the fiber is confined latitudinally within specific physical tracks (called "bands of polarization"). Figure 2.17 shows the spatial evolution in the Poincaré sphere of five different SOPs in their own "bands of polarization"; the simulated twisted fiber has a twist rate of 6.8 rad/m (~1 turn/m).

A fundamental benefit of this system consists in the reduced complexity of the receiver. Its simplicity derives from the simple need to detect only the component S_3 of the received SOP. Therefore, there is no need of the birefringence tracking circuit.

Starting from the simulative results described in the previous paragraph, it has been analyzed the statistical behavior of the different transmitted SOPs during their propagation along a twisted fiber [52]. All the statistical results have been obtained using the model and the physical parameters reported in [45]. In order to achieve statistically significant results, the simulations have been repeated 500 times (cycles), for each value of twist and for each distance of propagation.

The five transmitted symbols (Figure 2.17) have been chosen in such a way that the relative "bands of polarization" were symmetrical around the starting value of the latitude. To achieve this objective, we analyzed the

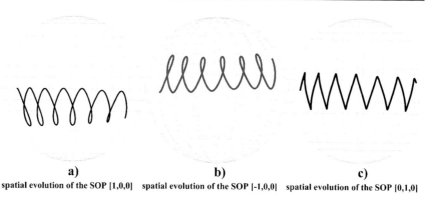

a)

spatial evolution of the SOP [1,0,0]

b)

spatial evolution of the SOP [-1,0,0]

c)

spatial evolution of the SOP [0,1,0]

Figure 2.19 Spatial evolution of different linear SOPs: a) SOP [1,0,0]; b) SOP [-1,0,0]; c) SOP [0,1,0].

behavior of different SOPs that belonged to the same band of polarization; the chosen test band was the equatorial band that includes the linear polarizations. Figure 2.19 shows the spatial evolution of different linear polarizations. It can be seen as the cycloidal spatial trajectory of the SOP is not symmetrical with respect to the equatorial plane if the starting value of S_2 is null, facing downward if S_1 is positive (Figure 2.19a) and upward in the opposite case (Figure 2.19b); conversely, the trajectory is symmetrical with respect to the equatorial plane if the starting value of S_2 is equal to one (Figure 2.19c). Moreover, the trajectories in Figure 2.19a and b are prolate cycloids while that in Figure 2.19c is an ordinary cycloid.

To enhance the visualization of the cycloidal patterns, we chose a weak twist rate of 1.5 rad/m. In fact, with a low twist rate, the spatial trajectory is a prolate cycloid, while increasing the twist rate it becomes first an ordinary cycloid and then a curtate cycloid (see also Figure 2.14). The same behavior holds true for the elliptical polarizations. On the contrary, circular polarization is flattened toward the pole because of the presence of strong spatial constraints (Figure 2.20).

The principal purpose of the statistical analysis was the study of the dependence of the spatial evolution of the SOP on the distance propagation and on the strength of the twisting process.

First, we analyzed the dependency of the transmitted SOP spatial evolution from the propagation distance of the optical field along the fiber. For each simulation cycle, it has been calculated the probability density function (pdf) of the third Stokes vector component S_3 relative to the transmitted symbols.

In fact, as demonstrated by (2.31)–(2.32), S_3 depends directly from the latitude angle and its variance is closely related to the width of its associated "band" (Figure 2.17).

Afterward, in order to consider all the cycle contributions, we derived the average pdf as the mean curve of all the executed simulations. In Figure 2.21, it is shown the behavior of the above-described mean pdf for different propagation distances.

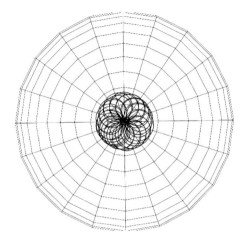

Figure 2.20 Top view of spatial evolution of a circular SOP.

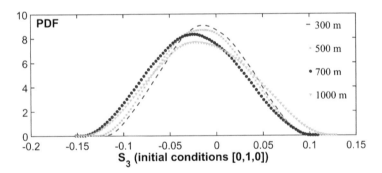

Figure 2.21 Mean S_3 pdf for different distances.

These plotted functions were obtained with different fiber distances but with the same value of the twist rate (6 rad/m). The transmitted symbol had a 45° linear polarization SOP with a Stokes vector equal to [0,1,0]. S_3 variance, and consequently the width of the bands of polarization, widens with increasing the distance of propagation. This variance growth has a linear dependence on the propagation distance as described in Figure 2.22, which shows a comparison between simulated values and a linearly fitted curve.

The statistical analysis then continued to understand how the strength of the twisting process could affect the trajectory of the SOP during its spatial evolution along the optical fiber. In particular, Figure 2.23 shows the dependency of S_3 pdf from different values of the twisting process. These plotted functions were obtained by considering a fixed value of the propagation distance equal to 500 m. Also, in this case, the transmitted symbol had a 45° linear polarization SOP with a Stokes vector equal to [0,1,0]. S_3 variance,

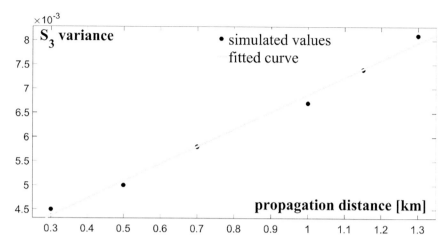

Figure 2.22 S_3 variance versus the propagation distance.

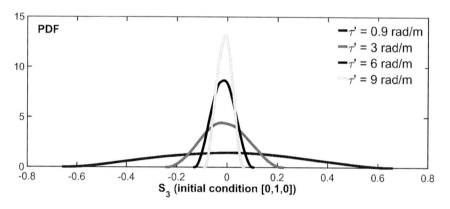

Figure 2.23 Mean S_3 pdf for different twisting values.

and consequently the width of the bands of polarization, narrows with the increasing twisting value. Therefore, an increase of the twisting process generates a potential throughput rise, with the same available bandwidth, for this type of multilevel polarization modulation, because it allows the presence of a greater number of bands (and, consequently symbols) on the Poincaré sphere.

This variance decrease has an exponential dependence from the twist rate, as shown in Figure 2.24, which shows a comparison between the simulated values and an exponentially fitted curve. Therefore, the statistical results prove how the width of the polarization bands has a dependence on the twisting process that is much stronger (exponential) than that from the propagation distance (linear).

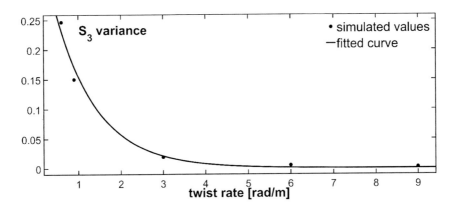

Figure 2.24 S₃ variance versus the twist rate.

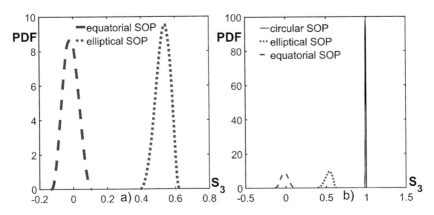

Figure 2.25 Mean S₃ pdf: a) comparison between equatorial and elliptical SOPs, b) comparison between equatorial, elliptical and circular SOPs.

The data originating from these simulations show that the twisting process gives rise to a physical track even tighter for circular polarization than for those equatorial and elliptical. Considering the same values of distance (500 m) and twist rate (6 rad/m), Figure 2.25a shows the comparison between a linear and an elliptical SOP, while in Figure 2.25b a circular SOP is added. It is clear Figure 2.25b, how large is the difference between pdf curves of circular polarization on one side and those of the equatorial and elliptical polarizations on the other side.

Another important result that can be deduced by Figure 2.25a and b is that the width of the "bands" decreases, proceeding from the equator to the pole. This behavior proves how a transmitted circular SOP is physically advantaged with respect to the others SOPs, in terms of a less probable deviation from its initial position, during the spatial propagation in a twisted optical fiber. It is reasonable to assume that this behavior is determined by the greater strength of the circular polarization with respect to the symmetry, also circular, of the fiber core.

2.4.4 Mathematical Model

The statistical analysis performed by simulation (described in 2.82) can be matched with a mathematical model proposed by Perrin [53], which characterizes the Brownian motion of a particle on the surface of a sphere. Perrin showed how the pdf of the colatitude angle θ as a function of time is given by:

$$f(\theta,t) = \sum_{k=0}^{\infty} \frac{1}{4\pi}(2k+1)e^{-k(k+1)Rt}\, P_k\left(\cos\theta\right) \tag{2.83}$$

where P_k is the Legendre polynomial of order k and R is the rotational diffusion coefficient (in rad^2/s). This function is obtained by starting from the initial conditions $f(0,0) = \infty$ and $f(\theta,0) = 0$ (initial conditions equivalent to a circular polarization). The pdf of Perrin has been manipulated using the methods exposed in [44] in order to calculate the pdf of $S_3 = \cos\theta$, as a function of a random variable. Moreover, Equation (2.83) has been adapted to the various initial conditions that correspond to the SOP of the transmitted symbols. For example, for the linear polarizations (colatitude equal to $\pi/2$) it has been set $f(\pi/2,0) = \infty$ and $f(\theta,0) = 0$. In order to make a valid comparison between the simulated model and the mathematical model, we adopted the following assumption: it is not possible to use, in this case, the rotational diffusion coefficient R indicated by Perrin, so it has been deduced by simulation results through the diffusion equation:

$$\frac{\left\langle\theta^2\right\rangle}{2} = R\,t \tag{2.84}$$

in which the simulative values of $<\theta^2>$ were used. The term $<\theta^2>$ represents the mean-square angular displacement in time t.

An important result is that for the linear and elliptical polarizations, this mathematical model represents an upper limit (the dotted envelope in Figure 2.26) with respect to the pdf curves of S_3.

In the case of circular polarization, the mathematical model has an asymptotic behavior (the dotted curve in Figure 2.27) with respect to the pdf curves of S_3.

Simulated values in Figure 2.26 and Figure 2.27 were obtained for a fixed distance of 300 m and a constant twist rate of 6 rad/m. The pdfs of Figure 2.26 and Figure 2.27 were obtained by adopting another software that was more suitable for simulating the behavior of very complex mathematical functions such as that described by the Equation (2.83). In particular, the adopted software was *Mathematica 10* [54].

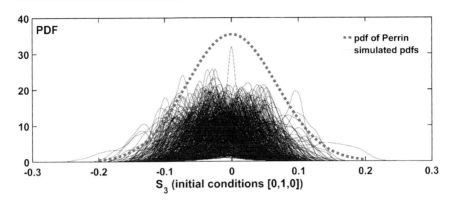

Figure 2.26 Perrin's pdf and simulated pdfs for a linear SOP.

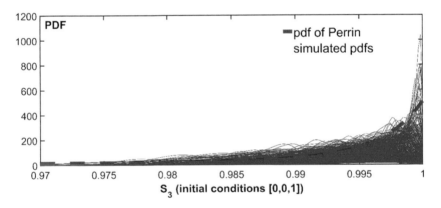

Figure 2.27 Perrin's pdf and simulated pdfs for a circular SOP.

A general result is that the diffusion process studied by Perrin flattens (with a trend toward uniform distribution) the pdf of the colatitude with increasing time as well as increasing distance (Figure 2.21). The decrease of the twist rate instead flattens the pdf obtained from the simulation results (Figure 2.23). It seems that the constraint introduced by the twist process is able to create privileged physical channels for the spatial evolution of the transmitted SOPs. These channels resist as long as the twist constraint is stronger than the natural process of diffusion.

2.4.5 Experimental Results

An experimental activity was carried out at the "Optical Communications Laboratory" of the Istituto Superiore delle Comunicazioni e delle Tecnologie dell'Informazione (ISCOM), an Institute that operates under the Ministry of Economic Development.

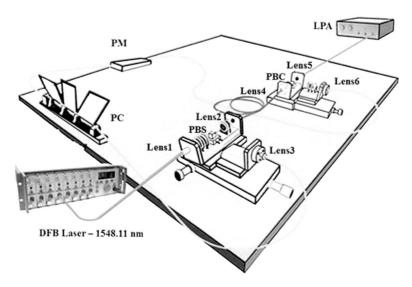

Figure 2.28 Graphical scheme of the realized ASPSK transmitter.

A prototype of Antipodal Stokes Parameters Shift Keying (ASPSK) transmitter was realized and developed on the basis of the scheme proposed in [55]. The realization of such ASPSK transmitter is a preparatory step to the development of a complete M-PolSK transmitter.

As shown in Figure 2.28, two different branches were realized: the direct (blue path) and the reflection one (yellow path). The setup consists of a DFB Laser, four fiber port collimators (Lenses 2/3/4/6) and two fixed benches SSMF (Lenses 1/5), a Lightwave Polarization Analyzer (LPA), Polarization Beam Splitter (PBS) and Polarization Beam Combiner (PBC) and two Polarization Maintaining Fibers (PMFs).

The presence on the reflection branch of a Phase Modulator (PM) and Polarization Controller (PC) acts directly on the component linearly polarized along $\hat{y}$, in order to realize a phase shift of 0° or 180°. In Figure 2.29 it is shown a picture of a section of the experimental setup, namely that related to the splitting and recombination of the emitted optical signal.

The transfer function of the PM observed on the oscilloscope is shown in Figure 2.30.

The optical signal is emitted from the laser source at a wavelength of 1548.11 nm and with an output power of 13 dBm. This signal is linearly polarized at 45°. The PBS splits the signal into the two components polarized along $\hat{x}$ and along $\hat{y}$.

The horizontally polarized component of the signal is collimated in the Lens2 (Figure 2.28), but before, it crosses two retarders ($\lambda/2$ and $\lambda/4$) in order to align the polarization of the signal along the slow or the fast axis of the PMF (the blue branch in Figure 2.28). On the other side, the vertically polarized signal component is sent to the PM after passing through the PC,

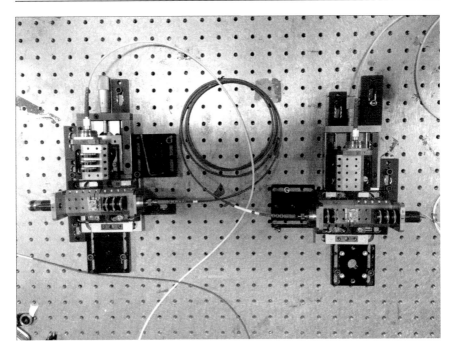

Figure 2.29 Section of the experimental set-up.

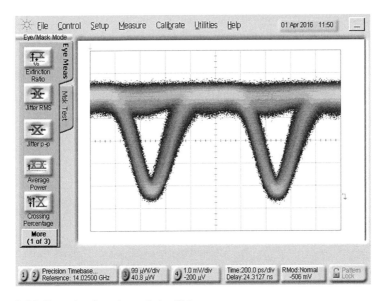

Figure 2.30 Transfer function of the PM.

necessary since it is used a SSMF; in this way it is recovered the correct polarization input to the PM. The following figures show the direct and the reflected components of the optical signal. Figure 2.31 and Figure 2.32 show the values assumed by the single components (direct and reflected, respectively).

The polarization analyzer (LPA) should show on the Poincaré sphere a "swing" movement between the points B and -B of the state of polarization of the signal, due to the recombination of the optical carrier polarized along $\hat{x}$ (the direct branch of Figure 2.28) and the modulating component polarized along $\hat{y}$ (the reflected branch of Figure 2.28) with a phase shift of 0° or 180°. In reality, after the recombination of the two components carried out by the PBC, it occurs what the Figure 2.33 shows.

It is clear how the output SOP is variable over time, because of the uniform birefringence introduced by the two PMFs used in the set-up. But the fundamental result that guarantees the proper behavior of the realized ASPSK transmitter is the presence of two antipodal regions that describes how a polarization modulation works. These regions are located near the equator of the sphere, describing a linear polarization, though they are slightly shifted angularly with respect to the points that define a linear polarization at 45° or at −45°. However, this problem is due to the instability of the devices used and the difficulty in aligning optical beams, due to the absence of micrometric controls. Figure 2.33 shows how the Degree Of Polarization (DOP) or P_D, as defined in (2.30), is reduced compared to the value assumed by the single components (direct and reflected in Figure 2.31 and Figure 2.32); this is caused by the receiver bandwidth of 1 kHz, which generates an average process of the state of polarization of the received samples. Table 2.3 summarizes the results of such a configuration.

Moreover, in order to verify the correct behavior of the realized transmitter, it is useful to perform an analysis of the output signal with the oscilloscope. Since the power level of the output signal is quite low, to display it on

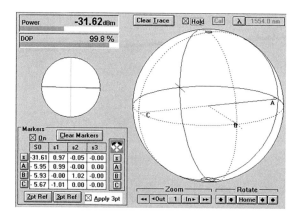

Figure 2.31 Direct component (linearly polarized along the $\hat{x}$ axis).

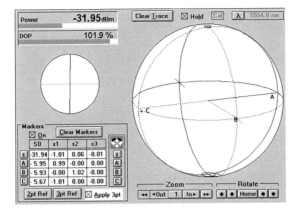

Figure 2.32 Reflected component (linearly polarized along the $\hat{y}$ axis).

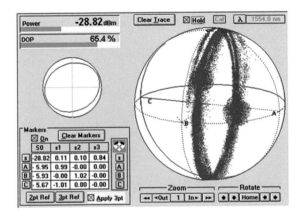

Figure 2.33 Optical signal after the PBC.

Table 2.3 Power levels of the optical signal

Polarized components	Power levels
Direct component (linearly polarized along $\hat{x}$)	−31.62 dBm
Reflected component (linearly polarized along $\hat{y}$)	−31.95 dBm
Recombined signal with polarization variable over time	−28.82 dBm

the oscilloscope present in the laboratory, it has been introduced an EDFA followed by a 1550 nm filter. Figure 2.34 shows the final output of the realized ASPSK transmitter.

As expected, the diagram has the same trend of the transfer function of the phase modulator (Figure 2.30); indeed, such behavior is the direct

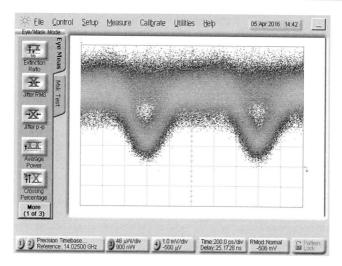

Figure 2.34 Final output of the ASPSK transmitter.

consequence of the overlap between the carrier, which is represented on the oscilloscope by a constant level, and the modulated signal.

2.4.6 Performance Evaluation

As described above, the circular birefringence, induced by the twisting process, defines the physical bands. In order to analyze the system performances, these bands, which constitute the decision regions, have been chosen after an optimization process. This performance evaluation starts from the considerations, described in [51], about the optimal decision regions for M-PolSK modulations as constellations of regular polyhedra inscribed in the Poincaré sphere. In particular, in [51] the decision regions, different for the various M-PolSK modulations, coincide with regular spherical polygons identified by all the spherical coordinates.

Nevertheless, the topological concept of the proposed modulation is completely different because it creates a "fluid" constellation of symbols belonging to physical bands of polarization and no longer anchored to rigid geometrical structures such as polyhedra. Therefore, only a spherical coordinate (colatitude) is involved in the definition of decision regions that, for this reason, present only a colatitude upper bound beyond which decision errors always happen. This limit is equal to $\pi/(M-1)$, where M is the number of symbols of the proposed system.

System performances for the "bands of polarization" modulation were calculated with respect to the statistical properties of two different noises: shot-noise and thermal noise. Previous sections of this book showed how the probability density functions are different depending on the transmitted SOP (narrower band of polarization and lower error probability for circular SOP). This result is valid until the signal propagates along the fiber. At the receiver side, other noise sources must be considered. As a matter of fact, birefringence effect is selective in its influence on SOP; on the contrary, noise processes like shot and thermal noise have the same behavior against all the SOPs. Under the hypothesis of independent equiprobable symbols, the symbol error probability can be written as:

$$P(e) = \frac{1}{M} \sum_{k=1}^{M} P\left(e|\vec{S_k}\right) = P\left(e|\vec{S_k}\right) \tag{2.85}$$

where $P\left(e|\vec{S_k}\right)$ is the error probability conditioned to the transmission of the symbol $\vec{S_k}$.

2.4.7 Coherent Detection Performance

In this case, we assumed the hypothesis of a receiver based only on the estimation of the Stokes parameter S_3 of the received optical field by means of a coherent optical front end. Therefore, the dominant noise source is the shot-noise. Coherent detection offers many benefits with respect to sensitivity, spectral efficiency and equalization potential [56, 57]. With the choice described above about the decision regions, only the colatitude θ is involved in the performance analysis. Starting from the results derived in [51] and after suitable manipulations due to the different considered decision regions, the conditional error probability can be expressed as:

$$P\left(e|\vec{S_k}\right) = 1 - F_\vartheta\left(\vartheta_1\right), \tag{2.86}$$

where $\theta_1 = \pi/(M-1)$ is the optimized colatitude upper bound of the decision regions and F_ϑ is the Cumulative Density Function (CDF). The CDF has the following expression [51]:

$$F_\vartheta(t) = 1 - \frac{1}{2} e^{-\frac{S_0}{4\sigma^2}(1-\cos t)} \left(1 + \cos t\right) \tag{2.87}$$

with t in the range $[0,\pi]$ and σ^2 the noise variance. The resulting error probability was analyzed in function of the signal-to-noise ratio per transmitted information bit η_b [51]:

$$\eta_b = \eta_s \frac{1}{\log_2 M} \tag{2.88}$$

where, the term η_s is equal to $T_0/2\sigma^2$, M is the number of symbols and σ^2 the noise variance.

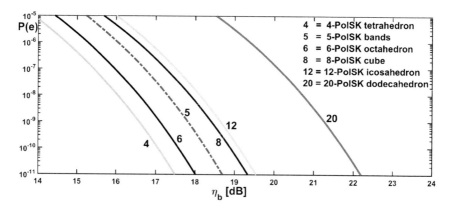

Figure 2.35 Comparison between M-PolSK [51] and 5-PolSK.

Figure 2.35 shows the comparison between the results reported in [51] and the performance relative to the proposed 5-PolSK bands polarization system.

The proposed system is located between 6-PolSK octahedron and 8-PolSK cube curves. It is slightly better (~0.1 dB) than the optimum 8-PolSK non-regular polyhedron described in [51] and not reported in Figure 2.35. The decision regions are calculated by means of a unique angle (colatitude), neglecting the longitudinal one. This choice from one side causes a penalty of less than 1 dB (~0.7 dB) with respect to the 6-PolSK octahedron, but from the other side it guarantees a simpler receiver based only on the knowledge of S_3 and with no birefringence tracking circuit.

2.4.8 Direct Detection Performance

In this case, we assumed the hypothesis of a receiver based only on the estimation of the Stokes parameter S_3 of the received optical field by means of a direct detection optical front end. Therefore, the dominant noise source is the receiver thermal noise. The objective was to compare the performances of the proposed system with the Polarization Modulated Direct Detection (PM-DD) systems described in [58]. In order to achieve a better comparison, the method used for the performance evaluation was the so-called union bound approximation utilized in [58] and described in [59] and [46]. As described in [47], this novel modulation model does not need fixed constellation symbols but only the knowledge of the associate bands of polarization. Nevertheless, in order to evaluate performance using the distance criterion, it is necessary the choice of a reference constellation of symbols. The chosen coordinates for the constellation symbols in the Stokes space are listed in Table 2.4.

Table 2.4 Symbols in the Stokes space

N	S_1	S_2	S_3
1	0	0	1
2	0	$\sqrt{2}/2$	$\sqrt{2}/2$
3	0	1	0
4	0	$\sqrt{2}/2$	$-\sqrt{2}/2$
5	0	0	-1

This choice fell on the symbols in Table 2.4 because it was considered the worst possible case with the minimum distance among all the nearest neighbors.

Distances between the symbols of the proposed system are reported in Table 2.5.

Under the hypothesis of independent equiprobable symbols, symbol error probability can be written as in (2.85). Conditional error probability has been derived as in [58], by using the same receiver parameters. System bit error probability P_{bit} was calculated as a function of symbol error probability P_{symbol} by means of the expression:

$$P_{bit} = \left(\overline{e} \cdot P_{symbol} \right) / \log_2 M, \qquad (2.89)$$

where M is the number of symbols and $\overline{e}$ is the average number of wrong bits in a wrong symbol [59]:

$$\overline{e} = \frac{1}{M-1} \sum_{k=1}^{\log_2 M} \binom{\log_2 M}{k} k = \frac{M \log_2 M}{2(M-1)}. \qquad (2.90)$$

Therefore, the bit error probability is given by:

$$P_{bit} = \frac{M}{2(M-1)} P_{symbol}. \qquad (2.91)$$

Table 2.5 Distance matrix for the 5-PolSK bands modulation

N	1	2	3	4	5
1	0	0.765	$\sqrt{2}$	1.847	2
2	0.765	0	0.765	$\sqrt{2}$	1.847
3	$\sqrt{2}$	0.765	0	0.765	$\sqrt{2}$
4	1.847	$\sqrt{2}$	0.765	0	0.765
5	2	1.847	$\sqrt{2}$	0.765	0

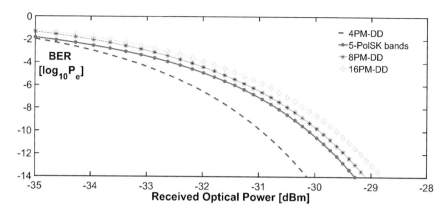

Figure 2.36 Comparison between PM-DD [58] and 5-PolSK.

The resulting bit error probability has been evaluated in function of the received optical power, measured in dBm.

Figure 2.36 shows the comparison between the results described in [58] and the evaluated performance of 5-PolSK bands polarization system in presence of thermal noise.

The proposed system is located between 4PM-DD and 8PM-DD curves, with a performance penalty of less than 1 dB (~0.9 dB) with respect to 4PM-DD.

2.5 MULTILEVEL COMBINED PHASE AND POLARIZATION SHIFT KEYING MODULATION IN TWISTED FIBERS

This paragraph illustrates, in detail, an innovative multilevel combined phase and polarization shift keying modulation in twisted fibers for LAN applications [60]. This modulation is able to extend the concept described in the previous paragraph. In particular, the proposed scheme will prove how the birefringence could be exploited also in four-dimensional Euclidean space in which it is possible to represent the electromagnetic wave with all its components (Betti *et al.* [16]).

The electromagnetic field is characterized by two quadratures in two polarization components; thus, in total, there are four Degrees Of Freedom (DOFs), which describe a four-dimensional signal space. The electric field can be written by means of the Jones column matrix [56]:

$$E = \begin{pmatrix} E_{x,r} + i\,E_{x,i} \\ E_{y,r} + i\,E_{y,i} \end{pmatrix} = \begin{pmatrix} |E_x|\exp(i\varphi_x) \\ |E_y|\exp(i\varphi_y) \end{pmatrix}, \tag{2.92}$$

where x and y represent the polarization components, and r and i the real and imaginary parts of the field, respectively. The phase terms φ_x and φ_y are defined in the interval $(-\pi,\pi]$. Another way to describe the electric field is to consider its phase, amplitude and polarization state (the latter being the relative phase and amplitude between the x and y components of the field) as [56]:

$$E = |E|\exp\left(i\varphi_a\right)J = |E|\exp\left(i\varphi_a\right)\begin{pmatrix} \cos\theta\exp\left(i\varphi_r\right) \\ \sin\theta\exp\left(i\varphi_r\right) \end{pmatrix}, \tag{2.93}$$

where $|E|^2 = |E_x|^2 + |E_y|^2$, $\theta = \sin^{-1}(E_y/|E|)$ and J denotes the Jones vector, which is usually normalized to unity. The term $\varphi_a = (\varphi_x + \varphi_y)/2$ represents the absolute phase of the field while the term $\varphi_r = (\varphi_x - \varphi_y)/2$ is the relative phase between the field vector components. The relative phase, defined in the interval $(-\pi,\pi]$, describes the ellipticity of the polarization state. The angle θ is usually indicated as the azimuth; it describes the orientation in the xy-plane of the linear polarization states or, more generally, the major axis of the polarization ellipse. The signal, with its four real components, can be expressed as [56]:

$$E = \begin{pmatrix} E_{x,r} \\ E_{x,i} \\ E_{y,r} \\ E_{y,i} \end{pmatrix} = \begin{pmatrix} |E|\cos\varphi_x \sin\theta \\ |E|\sin\varphi_x \sin\theta \\ |E|\cos\varphi_y \cos\theta \\ |E|\sin\varphi_y \cos\theta \end{pmatrix}. \tag{2.94}$$

The circular birefringence, induced by a suitable twisting process, is able to create predictable physical channels in which the polarization and the phase of the transmitted four-dimensional symbol are confined during the spatial propagation along the optical fiber [60]. This important result is well observable in the phase sphere (a descriptor of the polarization ellipse complementary to the Poincaré sphere and the Stokes parameters [61]).

In this way, a novel concept of modulation format can be realized by confining field's polarization and phase that can be used for hybrid modulation in predictable physical tracks [60].

By adopting this novel modulation scheme, it is possible to obtain the same performances of existent transmission systems but with the considerable advantage of a simpler receiver with a better cost-efficiency.

2.5.1 The Phase Sphere

Considering the electric field in its general definition E(r,t) as a solution of the vector wave equation (r(x,y,z) is the position and t the time). By Fourier analysis, an arbitrary vector wave may be expressed by means of two real vectors **p** and **q** as [62]:

$$E(r,t) = p(r)\cos\omega t + q(r)\sin\omega t = \text{Re}\{E(r)e^{-i\omega t}\},\qquad(2.95)$$

where **E** is the complex vector:

$$E(r) = p(r) + iq(r).\qquad(2.96)$$

From Equation 2.96, it is possible to derive the four real components of the electric field:

$$E = \begin{pmatrix} E_{x,r} \\ E_{x,i} \\ E_{y,r} \\ E_{y,i} \end{pmatrix} = \begin{pmatrix} p_x \\ q_x \\ p_y \\ q_y \end{pmatrix}.\qquad(2.97)$$

The Stokes parameters, defined in 2.33–2.35, can be rewritten in the following way [17]:

$$\begin{aligned} S_0 &= E_x E_x^* + E_y E_y^* = p_x^2 + q_x^2 + p_y^2 + q_y^2 \\ S_1 &= E_x E_x^* - E_y E_y^* = p_x^2 + q_x^2 - p_y^2 - q_y^2 \\ S_2 &= E_x E_y^* + E_x^* E_y = 2(p_x p_y + q_x q_y) \\ S_3 &= i(E_x E_y^* - E_x^* E_y) = 2(p_x q_y - q_x p_y). \end{aligned}\qquad(2.98)$$

As mentioned above, the Stokes parameters are independent from the absolute phase of the carrier. Therefore, the information about the phase is not present in the Stokes parameters, which only provide information about intensity, handedness, eccentricity and orientation (which are measurable quantities). But in a four-dimensional space, the phase can be obtained, as proved in [62] where it is denoted with the term ε, according to the relationship:

$$\tan 2\varepsilon = \frac{2\,p \cdot q}{p^2 - q^2}.\qquad(2.99)$$

This phase term is used in literature to investigate polarization singularities [61, 63].

Starting from 2.99, it is possible to introduce a phase sphere, complementary to the Poincaré sphere, by using as azimuth the phase ε rather than the orientation angle ψ [61] (see the polarization ellipse in Figure 2.37).

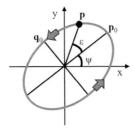

Figure 2.37 Polarization ellipse.

As the Stokes parameters characterize the Poincaré sphere, the phase sphere is described by the following phase parameters [61]:

$$
\begin{aligned}
T_0 &= p_x^2 + q_x^2 + p_y^2 + q_y^2 = S_0 \\
T_1 &= p^2 - q^2 = p_x^2 + p_y^2 - q_x^2 - q_y^2 \\
T_2 &= 2(\mathbf{p}\cdot\mathbf{q}) = 2(p_x q_x + p_y q_y) \\
T_3 &= S_3 = 2(p_x q_y - q_x p_y).
\end{aligned}
\tag{2.100}
$$

The zero-th and third phase parameters coincide to the corresponding Stokes parameters; therefore, the eccentricity dependence on latitude is the same as on the Poincaré sphere. It is evident how $\tan 2\varepsilon = T_2/T_1$ and $\tan 2\psi = S_2/S_1$ define the azimuth respectively in the phase sphere and in the Poincaré sphere. Therefore, the phase varies with azimuth on the phase sphere.

The Poincaré sphere corresponds to the image of the three-sphere by the Hopf map [64]. The Hopf circles correspond to the points with different phases ε. The phase sphere is similar, but the Hopf circles are parametrized by ψ and the two different spheres are correlated by changing p_y and q_x in the four-dimensional space [61].

2.5.2 A Novel Model in Twisted Fibers

Starting from the scheme proposed in the paragraph 2.4 and from the theory exposed above, a novel Multilevel Combined Phase and Polarization Shift Keying Modulation in twisted fibers in local environments was developed. The relative constellation is realized in the four-dimensional space to exploit both the polarization and the phase of the transmitted symbol.

2.5.2.1 Transmitter

The transmitter block can be derived easily from that described in [16]. In particular, the transmitter must be able to provide at its output an arbitrary electrical field, with given power P, represented by a generic vector E as shown in 2.94. It can be obtained from a continuous wave linearly polarized field by simultaneously modulating the phase and the polarization of the field; in this way the transmitter can be designed as a cascade of a phase and a polarization modulator [16]. This transmitter is able to build the desired constellation in the phase sphere by exploiting the relations in 2.100.

2.5.2.2 Propagation in a Twisted Fiber

The propagation of the four real components of the signal is governed by the wave equation as expressed in [17]:

$$\frac{d\begin{pmatrix} E_x \\ E_y \end{pmatrix}}{dz} + \frac{1}{2}i\beta\sigma\begin{pmatrix} E_x \\ E_y \end{pmatrix} = 0, \tag{2.101}$$

where $\boldsymbol{\beta} = (\beta_1, \beta_2, \beta_3)$ is the local birefringence vector and $\boldsymbol{\sigma} = (\sigma_1, \sigma_2, \sigma_3)$ the Pauli spin vector in the Stokes space. From 2.101 it is possible to derive the four differential equations that have been simulated by the software *MATLAB*:

$$\begin{aligned}
\frac{dE_{x,r}}{dz} &= \frac{1}{2}\left(\beta_1 E_{x,i} + \beta_2 E_{y,i} - \beta_3 E_{y,r}\right) \\
\frac{dE_{x,i}}{dz} &= -\frac{1}{2}\left(\beta_1 E_{x,r} + \beta_2 E_{y,r} + \beta_3 E_{y,i}\right) \\
\frac{dE_{y,r}}{dz} &= \frac{1}{2}\left(-\beta_1 E_{y,i} + \beta_2 E_{x,i} + \beta_3 E_{x,r}\right) \\
\frac{dE_{y,i}}{dz} &= \frac{1}{2}\left(\beta_1 E_{y,r} - \beta_2 E_{x,r} + \beta_3 E_{x,i}\right).
\end{aligned} \tag{2.102}$$

By adopting the Wai-Menyuk Model (WMM) [41], the components β_1 and β_2 can be considered as independent Langevin processes. By applying a suitable twisting process, the component β_3 becomes proportional to the twist rate [65]. The local birefringence vector $\boldsymbol{\beta}$ can be written as [45]:

$$\beta(z) = T(z)\begin{pmatrix} \beta_1(z) \\ \beta_2(z) \\ g\tau'(z) \end{pmatrix} \tag{2.103}$$

with $T(z)$ that indicates the rotation matrix of the cross-sectional plane z generated by the twisting process [45]:

$$T(z) = \begin{pmatrix} \cos 2\tau(z) & -\sin 2\tau(z) & 0 \\ \sin 2\tau(z) & \cos 2\tau(z) & 0 \\ 0 & 0 & 1 \end{pmatrix} \tag{2.104}$$

while $\tau(z)$ is the twist expressed in radians and $\tau'(z)$ is the twist rate expressed in rad/m. The parameter g is the stress-optic coefficient, which is related to optical fiber coupling parameters and represents the proportionality coefficient between twist rate and induced circular birefringence ($\beta_3(z) = g\tau'(z)$). A typical value of this parameter is $g \cong 0.14$ [45].

Considering to transmit the four-dimensional symbol [1,0,0,0] in a twisted fiber with a constant twist rate of 15 rad/m ($\cong$ 2.38 turns/m) and a length of 1 Km. The symbol is mapped (see Equation (2.100)) into the point [1,0,0] on the phase sphere and into the point [1,0,0] (see Equation (2.98)) on the Poincaré sphere.

Figure 2.38 shows how the twisting process is able to create, in the phase sphere, predictable physical channels for all the three phase parameters (T_1, T_2, $T_3=S_3$) that can be derived from the Equation (2.100) by mapping the transmitted four-dimensional symbol. The phase parameters, during the spatial propagation along the optical twisted fiber, slightly fluctuate around their initial value (this oscillation decreases with the rise of the twisting process) thanks to the induced circular birefringence. Conversely, the Stokes parameters S_1 and S_2 assume all the possible values in the range [−1,1]. The S_3 parameter is the common term between the two different spheres. The confined trajectories of T_1 and T_2 prove how the twisting process physically constrains the phase of the field as well as the trajectory of S_3 proves how the twisting process constrains the latitude of the polarization.

This evidence induces an important consideration: in the model described in the paragraph 2.4, while requiring a simpler receiver with performances comparable to existing modulations, the number of transmitted symbols (the single channel throughput) was limited by the number of bands that the twisting process could create. In fact, for each band of polarization only one symbol could be transmitted. This novel modulation removes this flaw because in

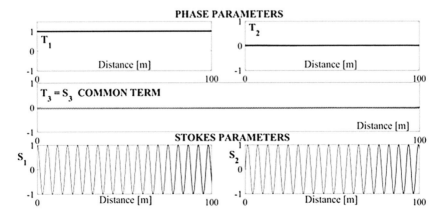

Figure 2.38 Phase and Stokes parameters of the symbol [1,0,0,0] (the figure shows only the first 100 meters, since the trajectories remain unchanged).

the phase sphere all the three parameters are confined by the twisting process; in this way it is possible to increase the number of symbols, and consequently the single channel throughput, as in the constellations shown in [51].

Figures 2.39 and 2.40 highlight the difference in the spatial propagation of the same four-dimensional transmitted symbol projected in the two

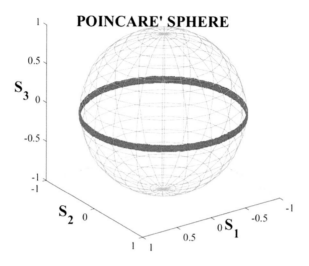

Figure 2.39 Spatial propagation of [1,0,0,0] in the Poincaré sphere.

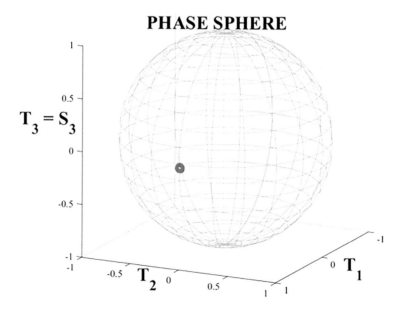

Figure 2.40 Spatial propagation of [1,0,0,0] in the phase sphere.

different spheres (the Poincaré sphere and the phase sphere). As described above, it is clear how a suitable twisting process is able to totally confine (longitudinally with the two terms related to the phase T_1 and T_2 and latitudinally with the term S_3 related to the polarization), in the phase sphere, the projection of the transmitted symbol that slightly fluctuates around its initial value. Conversely, in the Poincaré sphere, the constraint introduced by the twisting process is only related to the latitude of the polarization.

The fundamental result is the proof that, in the phase sphere, the induced circular birefringence is able to confine physically both the polarization and the phase of the electromagnetic field during the spatial propagation of the symbol along the twisted optical fiber.

2.5.2.3 Receiver

The block scheme of the receiver is shown in Figure 2.41. The received field polarization components are divided by using a polarization beam splitter; then they are coupled with the corresponding polarization components of the local oscillator by means of two balanced $\pi/2$ optical hybrids [16]. The four photodiodes detect the four fields at the output of the hybrids; the obtained electrical signals are proportional to the components of the representative vector of the received field [16]. After detection, there are four identical IF filters. This first part of the receiver, which includes the optical and the IF stages, is the same of the receiver described in [16].

The second part of the receiver is the original one. The objective of this section of the receiver is to derive the three phase parameters, starting from the four real components of the received field, as described in Equation (2.100).

A comparison between the receiver scheme of Figure 2.41 and those described in the various articles ([16, 51, 10] and [66]) about Multilevel-PolSK (M-PolSK) immediately shows the main difference: the absence in

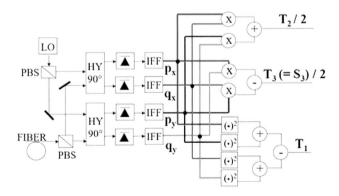

Figure 2.41 Receiver scheme.

the proposed receiver of the birefringence tracking circuit. Such a simplification, with no penalties about the performance as illustrated in the next section, obtained by exploiting a natural feature of the optical fiber such as the birefringence, guarantees an evident improvement of the costs/benefits ratio.

2.5.3 System Performance Evaluation

The system performance can be calculated starting from the statistical characterization of the noise processes that alter the estimation of the transmitted field. We assume that the signal at the output of the IF filters in Figure 2.41 is perturbed by shot and receiver noise, assumed as additive white Gaussian noise. Furthermore, we consider that the transmitted symbols are equiprobable and equipower.

For this purpose, we choose to build in the phase sphere the same constellations of symbols studied in [51] in the Poincaré sphere. These constellations may be constituted by regular polyhedra, in which the different symbols are the vertices, inscribed in the phase sphere or by optimum constellations that do not coincide with any geometric figure but that are built by means of a maximization of the minimum distance among the symbols in the phase sphere. The problem of how to find the arrangement of M points on a unit sphere that maximizes the minimum distance between any two points is known as the Tammes's Problem [67]. For M = 4,6,12 the points corresponding to the vertices of a regular polyhedron represent the optimal choice; however, for M = 8 this is no longer true (the optimal figure is the square antiprism and not the cube inscribed in the sphere).

The method used in [51] is the Maximum Likelihood (ML) decision rule, which permits to draw the ML decision regions for a generic signal constellation. In the case of the regular polyhedra, the ML decision regions are congruent (with the same shape and size) and their intersections with the phase sphere are bounded by regular spherical polygons [51]. In the case of the optimum constellations the decision regions turn out to be congruent, although they may not be regular spherical polygons.

Thanks to the suitable twisting process applied in the proposed system, there is no need of a birefringence tracking circuit. The transmitted symbol, during the propagation in the twisted fiber, oscillates, in the phase sphere, along a circumference whose radius is negligible (with the adopted twist rate and a distance comparable with LAN environment) and whose center coincides with the symbol itself (Figure 2.40). The symbols on the phase sphere are thus subjected only to the noise introduced by the receiver. The symbol error probability is therefore obtained as in [51]. The optimum decision rule, adopted with the aim to minimize the average symbol error probability, is based on the application of the Maximum A Posteriori (MAP) method that, when the transmitted symbols are equally likely, becomes equivalent to the ML rule [51].

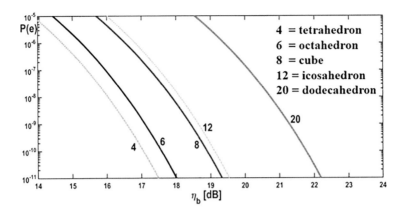

Figure 2.42 Symbol error probability for regular polyhedra.

Figure 2.42 shows the symbol error probability P(e) as a function of the signal-to-noise ratio per transmitted information bit η_b for the regular polyhedron (M=4,6,8,12,20) inscribed within the phase sphere. The term η_b can be expressed with the relation 2.88.

Also, the performance for optimum constellations (with M=8,16,32) are the same of [51].

2.5.4 Considerations

The proposed model is an innovative multilevel combined polarization and phase shift keying modulation for LAN applications.

The main result is the prove that, by exploiting in the optical fiber the circular birefringence, induced by a suitable twisting process, it is possible to confine, within predictable physical channels, the polarization and the phase of the transmitted field.

This confinement takes place in the phase sphere, a descriptor of the polarization ellipse complementary to the Poincaré sphere and the Stokes parameters.

The phase sphere becomes a powerful way to highlights the physical constraints imposed by the twisting process on the polarization and the phase of the field during its propagation along the twisted fiber.

The proposed system achieves the same performance of the standard M-PolSK modulations, but it presents the great advantage to require a receiver much less complex. This lower complexity is due to the absence of tracking birefringence circuit that in standard M-PolSK systems are necessary to compensate SOP fluctuations.

Therefore, this system is able to provide the same single channel throughput of traditional M-PolSK modulations by achieving a lower implementation penalty.

REFERENCES

[1] J. Tyndall, "On some phenomena connected with the motion of liquids," *Proceedings of the Royal Institution of Great Britain*, pp.1–446, 1854.

[2] J. L. Baird, "An improved method of and means for producing optical images". *British Patent Patent*, vol.285, p.738, 1927.

[3] C. W. Hansell, Patent U.S. Patent 1,751,584, 1930.

[4] H. Lamm, Patent Z. Instrumentenk. 50, 579, 1930.

[5] N. S. Kapany, *Fiber Optics: Principles and Applications*, San Diego, CA: Academic Press, 1967.

[6] F. P. Kapron, D. B. Keck and R. D. Maurer, "Radiation losses in glass optical waveguides," *Applied Physics Letters*, vol.17, p.423, 1970.

[7] T. Miya, Y. Terunuma, T. Hosaka and T. Miyoshita, "Ultimate low-loss single-mode fiber at 1.55 μm," *Electronics Letters*, vol.15, p.106, 1979.

[8] A. Snyder and J. D. Love, *Optical Waveguide Theory*, London: Chapman & Hall, 1983.

[9] S. C. Rashleigh, "Origins and control of polarization effects in single-mode fibers," *IEEE Journal of Lightwave Technology*, vol.LT1, no.2, pp.312–330, 1983.

[10] S. Betti, F. Curti, G. De Marchis and E. Iannone, "Multilevel coherent optical system based on Stokes parameters modulation," *IEEE Journal of Lightwave Technology*, vol.8, pp.1127–1136, 1990.

[11] G. Stokes, "On the composition and resolution of streams of polarized light from different sources," *Transactions of the Cambridge Philosophical Society*, vol.9, pp.399–416, 1852.

[12] H. Poincaré, "Chap. 12," in *Théorie Mathematique de la Lumieré*, vol.2, Paris, Gauthiers-Villars, 1892.

[13] H. Mueller, "Memorandum on the polarization optics of the photo-elastic shutter," in Report No. 2 of OSRD, project OEMsr-567, 1943.

[14] H. Mueller, "The foundation of optics," *Journal of the Optical Society of America*, vol.38, p.661, 1948.

[15] R. E. Schuh, *Characterization of Birefringence in Optical Fibers*, Department of Electronic Systems Engineering - University of Essex, 1997.

[16] S. Betti, F. Curti, G. De Marchis and E. Iannone, "A novel multilevel coherent optical system: 4-Quadrature signaling," *IEEE Journal of Lightwave Technology*, vol.9, no.4, pp.514–523, 1991.

[17] J. P. Gordon and H. Kogelnik, "PMD fundamentals: Polarization mode dispersion in optical fibers," *PNAS*, vol.97, no.9, pp.4541–4550, 2000.

[18] R. Ulrich and A. Simon, "Polarization Optics of twisted single-mode fibers," *Applied Optics*, vol.18, no.13, pp.2241–2251, 1979.

[19] S. Personik, "Receiver Design for Digital Fiber Optic Communication Systems, Parts I and II," *Bell System Technical Journal*, vol.52, no.6, pp.843–886, 1973.

[20] C. Poole and R. Wagner, "Phenomenological approach to polarization dispersion in long single-mode fibres," *Electronics Letters*, vol.22, no.19, pp.1029–1030, 1986.

[21] H. Sunnerud, M. Karlsson, C. Xie and P. A. Andrekson, "Polarization mode dispersion in high-speed fiberoptic transmission systems," *IEEE Journal of Lightwave Technology*, vol.20, no.12, pp.2204–2219, 2002.

[22] M. Shtaif and A. Mecozzi, "Polarization-dependent loss and its effect on the signal-to-noise ratio in fiber-optic systems," *IEEE Photonics Technology Letters*, vol.16, no.2, pp.671–673, 2004.

[23] G. J. Foschini and C. D. Poole, "Statistical theory of polarization dispersion in single mode fibers," *IEEE Journal of Lightwave Technology*, vol.9, pp.1439–1456, 1991.

[24] G. P. Agrawal, *Fiber-Optic Communications Systems*, 3rd ed., John Wiley & Sons, 2002.

[25] F. Bruyere and O. Audouin, "Penalties in long-haul optical amplifier systems due to polarization dependent loss and gain," *IEEE Photonics Technology Letters*, vol.6, pp.654–656, 1994.

[26] E. Lichtman, "Limitations imposed by polarization-dependent gain and loss on all-optical ultralong communication systems," *IEEE Journal of Lightwave Technology*, vol.13, pp.906–913, 1995.

[27] D. N. Payne, A. J. Barlow and J. Ramskov Hansen, "Development of low and high birefringence optical fibers," *IEEE Journal of Quantum Electronics*, vol.18, pp.477–488, 1982.

[28] I. P. Kaminow and V. Ramaswamy, "Single-Polarization optical fibers: Slab model," *Applied Physics Letters*, vol.34, pp.268–270, 1979.

[29] W. Eickhoff, "Stress-induced single-polarization single-mode fiber," *Optics Letters*, vol.7, pp.629–631, 1982.

[30] J. Sakai and T. Kimura, "Birefringence caused by thermal stress in elliptically deformed core optical fibers," *IEEE Journal of Quantum Electronics*, vol.18, pp.1899–1909, 1982.

[31] R. Ulrich, S. C. Rashleigh and W. Eickhoff, "Bending-induced birefringence in single mode fiber," *Optics Letters*, vol.5, pp.273–275, 1980.

[32] S. C. Rashleigh and R. Ulrich, "High-birefringence in tension coiled single-mode fibers," *Optics Letters*, vol.5, pp.354–356, 1980.

[33] R. Ulrich and S. C. Rashleigh, "Polarization coupling in kinked single-mode fiber," *IEEE Journal of Quantum Electronics*, vol.18, pp.2032–2039, 1982.

[34] Y. Namihira, M. Kudo and Y. Mushiaka, "Effect of mechanical stress on the transmission characteristics of optical fibers," *Electronics and Communications in Japan*, vol.60C, pp.107–115, 1977.

[35] A. Kumar and R. Ulrich, "Birefringence of optical fiber pressed into a V-groove," *Optics Letters*, vol.6, pp.644–646, 1981.

[36] B. A. Saleh and M. C. Teich, *Fundamentals of Photonics*, New York: John Wiley & Sons, 1991.

[37] A. E. Craig and K. Chang, *Handbook of Optical Components and Engineering*, New Jersey: John Wiley & Sons, 2003.

[38] S. Huard, *Polarization of Light*, Paris: John Wiley & Sons, 1997.

[39] R. Waynant and M. Ediger, *Electro-Optics Handbook*, New York: McGraw-Hill Professional, 2000.

[40] A. Galtarossa, L. Palmieri, M. Schiano and T. Tambosso, "Statistical character-ization of fiber random birefringence," *Optics Letters*, vol.25, no.18, pp.1322–1324, 2000.

[41] P. K. A. Wai and C. R. Menyuk, "Polarization mode dispersion, decorrelation, and diffusion in optical fibers with randomly varying birefringence," *IEEE Journal of Lightwave Technology*, vol.14, pp.148–157, 1996.

[42] C. D. Poole, "Statistical treatment of polarization dispersion in single-mode fiber," *Optics Letters*, vol.13, no.8, pp.687–689, 1988.

[43] P. K. A. Wai and C. R. Menyuk, "Polarization decorrelation in optical fibers with randomly varying birefringence," *Optics Letters*, vol.19, no.19, pp.1517–1519, 1994.

[44] A. Papoulis, *Probability, Random Variables and Stochastic Processes*, 3rd ed., New York: McGraw-Hill, 1991.

[45] A. Galtarossa and L. Palmieri, "Measure of twist-induced circular birefringence in long single-mode fibers: theory and experiments," *IEEE Journal of Lightwave Technology*, vol.20, pp.1149–1159, 2002.

[46] A. Simon and R. Ulrich, "Evolution of polarization along a single mode fibre," *Applied Physics Letters*, vol.31, pp.517–520, 1977.

[47] P. Perrone, S. Betti and G. G. Rutigliano, "Optical communication system for high-capacity LAN," *Microwave and Optical Technology Letters*, vol.58, pp.389–393, 2016.

[48] P. Drexler and P. Fiala, "Optical fiber birefringence effects – Sources, utilization and methods of suppression (Chapter 7)," in *Recent Progress in Optical Fiber Research*, 2012, pp.127–150.

[49] [Online]. Available: https://it.mathworks.com/help/matlab/math/choose-an-ode-solver.html. [Accessed January 2017].

[50] D. Tentori and A. Garcia-Weidner, "Jones birefringence in twisted single-mode optical fibers," *Optics Express*, vol.21, no.26, pp.31725–31739, 2013.

[51] S. Benedetto and P. T. Poggiolini, "Multilevel polarization shift keying: Optimum receiver structure and performance evaluation," *IEEE Transactions on Communications*, vol.42, no.2/3/4, pp.1174–1186, 1994.

[52] P. Perrone, S. Betti and G. G. Rutigliano, "Statistical Model and Performance Analysis of a Novel Multilevel Polarization Modulation in Local 'Twisted' Fibers," *Photonics*, vol.4, no.1, 2017.

[53] F. Perrin, "Etude mathématique du mouvement Brownien de rotation," *Annales scientifiques de l'École Normale Supérieure*, vol.45, 1928, pp.1–51.

[54] [Online]. Available: http://www.wolfram.com/mathematica/?source=nav. [Accessed January 2017].

[55] S. Betti, F. Curti, G. De Marchis and E. Iannone, "Phase Noise and Polarization State Insensitive Optical Coherent Systems," *IEEE Journal of Lightwave Technology*, vol.8, no.5, pp.756–767, 1990.

[56] E. Agrell and M. Karlsson, "Power efficient modulation formats in coherent transmission systems," *IEEE Journal of Lightwave Technology*, vol.27, pp.5115–5126, 2009.

[57] P. J. Winzer and R. J. Essiambre, "Advanced optical modulation formats," *Proceedings of the IEEE*, vol.94, no.5, pp.952–985, 2006.

[58] S. Betti, G. De Marchis and E. Iannone, "Polarization Modulated Direct Detection Optical Transmission Systems," *IEEE Journal of Lightwave Technology*, vol.10, no.12, pp.1985–1997, 1992.

[59] J. G. Proakis, *Digital Communications*, 2nd ed., New York: McGraw-Hill, 1989.

[60] P. Perrone, S. Betti e G. Rutigliano, "A novel coherent multilevel combined phase and polarization shift keying modulation in twisted fibers," *Fiber and Integrated Optics*, vol.36, n. 4–5, pp.181–195, 2017.

[61] M. R. Dennis, "Polarization singularities in paraxial vector fields: morphology and statistics," *Optics Communications*, vol.213, pp.201–221, 2002.

[62] M. Born and E. Wolf, *Principles of Optics*, Oxford: Pergamon Press, 1959.

[63] J. F. Nye, "Lines of circular polarization in electromagnetic wave fields," *Proceedings of The Royal Society*, vol.A 389, pp.279–290, 1983.

[64] H. Hopf, "Uber die Abbildungen der dreidimensionalen Sphare auf die Kugelflache," *Mathematische Annalen*, vol.104, pp.637–665, 1931.

[65] R. Ulrich and A. Simon, "Polarization Optics of twisted single-mode fibers," *Applied Optics*, vol.18, pp.2241–2251, 1979.

[66] S. Benedetto, R. Gaudino and P. Poggiolini, "Polarization recovery in optical polarization shift-keying systems," *IEEE Transactions on Communications*, vol.45, pp.1269–1279, 1997.

[67] R. M. L. Tammes, "On the origin of number and arrangement of places of exit on the surface of pallen grains," *Recueil des Travaux Botaniques Neerlandais*, vol.27, pp.1–84, 1930.

Chapter 3

Multidimensional Optical Modulations

3.1 INTRODUCTION

One of the elements needed for information transmission is the code. The choice of the transmission code is closely linked to channel characteristics and to the signals used to transmit information through the channel but is also linked to some desired features of transmitter and receiver. A code is a set of code words or symbols that make up the alphabet with which information is represented. More precisely, referring to the model proposed by Shannon [1], the transmitter contains a code book with M different code words. Each code word is a sequence of n real numbers which can be regarded as a point $X\{x_1, x_2, ..x_n\}$ in a Euclidean space S of n dimensions.

Some characteristics of the codes are particularly important for the practical implementation of the transmission systems as they are associated with physical quantities used in the transmission of the signals. For example, the energy of a symbol $(X_i \cdot X_i)$ is associated with the power necessary to transmit that symbol and in a real system it is desirable to have equipotential symbols. Furthermore, the Euclidean distance between symbols determines the ability to distinguish them at the receiver in the presence of noise and therefore a code with distant symbols, possibly equally spaced, is desirable. For these reasons, already in the sixties Slepian [2] and Ottoson [3] concentrated their attention on codes with such characteristics, based on equal-energy signals which were defined as M points on a sphere in n-dimensional Euclidean space. The idea behind these codes was to start from one code word and obtain the others by permuting the components of the vector and replacing them with their negatives, hence the name of permutation modulation.

3.2 FOUR-DIMENSIONAL MODULATION

In 1974 Welti [4] resumed these works focusing his attention on the four-dimensional codes with the aim of using two separate two-dimensional (amplitude and phase) channels, showing that the 4-D codes allowed a gain

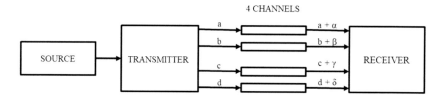

Figure 3.1 System configuration based on four independent noisy channels.

of 1 dB in terms of signal to noise ratio (E_b/N_0) for the same information transmitted. One of the numerous codes taken into consideration by Welti and indicated with C_2 and called "hypercube code" is a code consisting of 16 code words positioned in the vertices of a four-dimensional hypercube centered in the origin, already described and analyzed in [5] and that we will find later in the discussion. Figure 3.1 represents a four-channel communication system.

It is worth noting that the four communication channels are purely theoretical, not interfering with each other and that each channel is affected by an independent additive Gaussian white noise. The four independent channels can be practically implemented in a number of ways, provided that the non-interference requirements are respected and the noises are statistically independent. For example, you can use two passband channels each with separately modulated in-phase and quadrature-phase components.

After reviewing the most promising four-dimensional codes, the authors showed how the use of these, compared to a system based on two-dimensional codes, made it possible to transmit the same information with a transmission power lower than about 1dB. This apparently minimal difference can become important when the intensive use of optical fibers causes non-linear effects that reduce their capacity.

For a long time, four-dimensional codes have remained a mathematical artifice transposed into real systems using at most two-dimensional modulations (QAM, QPSK, etc.) consisting of the combination of one-dimensional techniques (ASK, PSK, etc.). The technological evolution of optical fibers, optical generators and amplifiers allowed an increase in performance satisfactory for the growing transmission needs.

A paradigm shift in this direction occurred in 1991 with the presentation of the 4-Quadrature Signaling [6–8] in which for the first time a four-dimensional code could be transmitted through four physical quantities of the same electromagnetic field, without resorting to the use of channels physically separated.

The authors started by the subsequent notation of the transversal components of the electromagnetic field propagating along the z direction:

$$E = E_x\mathbf{u}_x + E_y\mathbf{u}_y = \left(E_{xr} + iE_{xi}\right)\mathbf{u}_x + \left(E_{yr} + iE_{yi}\right)\mathbf{u}_y$$
$$= \left(a + ib\right)\mathbf{u}_x + \left(c + id\right)\mathbf{u}_y \tag{3.1}$$

where $\mathbf{u}_x$ and $\mathbf{u}_y$ are respectively the unit vectors along the x and y axis and:

$$
\begin{aligned}
E_{xr} &= \mathrm{Re}\left(E_x\right) = a \\
E_{xi} &= \mathrm{Im}\left(E_x\right) = b \\
E_{yr} &= \mathrm{Re}\left(E_y\right) = c \\
E_{yi} &= \mathrm{Im}\left(E_y\right) = d.
\end{aligned}
\tag{3.2}
$$

They observed that the four components (a, b, c, d) of the electromagnetic field could be determined by suitably modifying both the polarization and the phase of the electromagnetic field. To achieve this they devised a transmitter consisting of a phase modulator and a cascaded polarization modulator (Figure 3.2).

The encoder converts each element of the sequence m(t) into (α,β,γ) that drive the phase modulators in order to obtain an electromagnetic field with components (a, b, c, d).

Optical fiber propagation can be modeled, neglecting non-linear effects, with attenuation A, phase shift $B(\omega)$ and polarization response $\mathbf{J}$. The received optical signal can be expressed as:

$$
\begin{bmatrix} s_x \\ s_y \end{bmatrix} = e^{-\left[A - iB(\omega)\right]} \mathbf{JE} = e^{-\left[A - iB(\omega)\right]} \begin{bmatrix} u_1 & u_1 \\ -u_2^* & u_1^* \end{bmatrix} \begin{bmatrix} E_x \\ E_y \end{bmatrix}.
\tag{3.3}
$$

The vector S containing the four components of the received electromagnetic field can be related to the transmitted one:

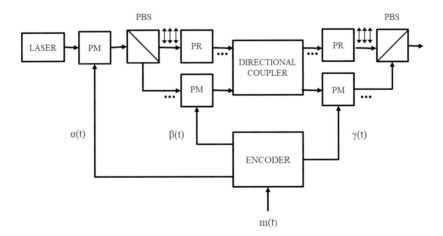

Figure 3.2 4-Quadrature transmitter scheme. PM: Phase Modulator, PBS: Polarization Beam Splitter, PR: Polarization Rotator of $\pi/2$.

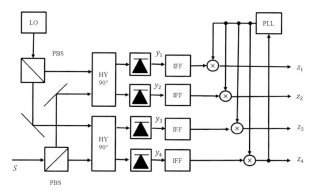

Figure 3.3 4-Quadrature Optical and IF receiver stages. LO: Local Oscillator, PBS: Polarization Beam Splitter, HY 90°: 90° balanced Hybrid, IFF: Intermediate Frequency Filter, PLL: Phase Locked Loop.

$$S = e^{-A}J_r E = e^{-A} \begin{bmatrix} u_{1R} & -u_{1I} & u_{2R} & -u_{2I} \\ u_{1I} & u_{1R} & u_{2I} & u_{2R} \\ -u_{2R} & -u_{2I} & u_{1R} & u_{1I} \\ u_{2I} & -u_{2R} & -u_{1I} & u_{1R} \end{bmatrix} \begin{bmatrix} a \\ b \\ c \\ d \end{bmatrix} \quad (3.4)$$

In order to reconstruct the transmitted signal, the receiver must compensate for the attenuation by amplification, but also for phase and above all polarization variations. The authors proposed two possible receivers which differ in baseband processing. The receiver common optical and IF stages are shown in Figure 3.3.

The optical input signal S is broken down into the four desired components which are measured by the four photodiodes. The signals thus obtained are filtered and subsequently demodulated by means of the PLL driven by the baseband circuit.

There are two different baseband processing schemes. Both have to compensate for the slow polarization fluctuations caused by the birefringence of the optical fiber and the more rapid phase fluctuations in order to generate the control signal for the PLL. The first scheme, illustrated in Figure 3.4, accomplishes these tasks by means of the inversion of the fiber real Jones matrix J_r.

The second scheme, illustrated in Figure 3.5, provides more complex algorithms and electronics and keeps updated a version of the modulation constellation modified taking into account the aforementioned fluctuations. The greater complexity is compensated by the greater flexibility of this scheme which can be applied to any constellation.

The authors showed higher spectral efficiency than other multilevel systems. The system did not show a great increase in performance but it paved the way for a new multilevel coding model, able to make the most of the peculiarities of the electromagnetic field that propagates in the optical fibers.

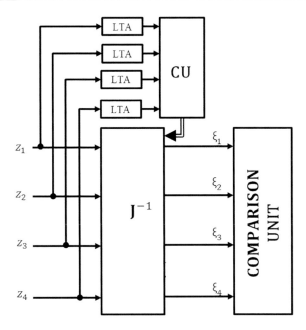

Figure 3.4 4-Quadrature band base processing by Jones matrix estimation. LTA: Long-Term Average, CU: Calculation Unit, J^{-1}: multiplier by the inverse Jones matrix.

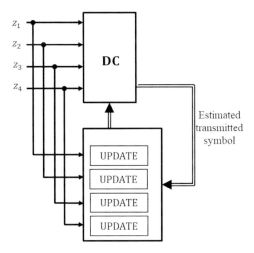

Figure 3.5 Baseband processing by reference vector. DC: Decision Circuit.

In particular, this system made it possible to exploit the four degrees of freedom offered by an electromagnetic field, the highest number of degrees of freedom used up to that time.

For two decades these results have not attracted much interest also because, as already mentioned, the technological development allowed a rapid increase in performance and the efforts were concentrated in this direction. When the rapidity of performance growth declined, the scientific community began to compare the various transmission techniques in search of optimizations in terms of spectral efficiency or power efficiency.

In 2009 Agrell and Karlsson [9] compared a series of multilevel modulation formats in two, three, four dimensions in search of an optimization of the transmission power which, as we know, affects non-linear phenomena. In particular, in this work the authors show how some modulation techniques are the evolution in a greater dimensionality of already existing techniques such as the transition from BPSK to QPSK to DP-QPSK. In addition, the authors resumed the works on theory of communications in four dimensions [4, 10–12] and those on coherent optical systems in four dimensions [6, 13–15] and came to establish a maximum advantage on sensitivity of 1.76 dB for the best four-dimensional transmission technique (PS-DPSK) compared to the best one-dimensional transmission technique (BPSK).

In 2012 the same authors [16] reviewed the known multilevel modulation techniques in order to identify the most efficient ones from the point of view of exploiting the optical spectrum, obtaining an advantage between 1.51 dB and 1.96 dB. Also in this case the gain is obtained by using multi-level modulations in four dimensions (SP-QAM) in which the arrangement of the symbols is optimized to make the most of the distances obtainable in a four-dimensional Euclidean space.

In 2014, Karlsson [17] showed that four is the maximum number of degrees of freedom usable according to the laws of quantum physics. In his work the author proposes a new formalism for the representation of the electromagnetic field that propagates in coherent optical communications, an alternative to the typically used 2×2 Jones matrices [18], described in Section 2.2.1. The new representation is based on rotations in the four-dimensional Euclidean space: the possible rotations are six. Four of these have a physical meaning and two are non-physical rotations, that is, they cannot occur because they do not obey the fundamental quantum mechanical boson commutation relations (Table 3.1).

Rotations are particular geometric transformations that leave some geometric places unchanged. Also, the matrices describing rotations have particular properties. For example, a four-dimensional rotation matrix belongs to group O_4, that is, it is an orthogonal matrix that multiplied by its transpose gives the identity matrix. It can be shown [17] that a rotation transformation can be expressed in the form of an exponential function whose exponent is a real skew-symmetric four-dimensional matrix.

Any four-dimensional rotation can be obtained as a linear combination of the six fundamental rotations that can occur in a four-dimensional space. Karlsson defined six matrices (Equations [3.5] and [3.7]) which form the basis of the possible rotations and are the generating matrices for four-dimensional rotations.

Table 3.1 Jones, Stokes–Mueller and four-dimensional
representations

Name	Formula
Jones	$E = \begin{pmatrix} E_x \\ E_y \end{pmatrix}$
Stokes–Mueller	$E = \begin{pmatrix} \lvert E_x \rvert^2 - \lvert E_y \rvert^2 \\ 2\,\mathrm{Re}\left(E_x E_y^*\right) \\ -2\,\mathrm{Im}\left(E_x E_y^*\right) \end{pmatrix}$
4-Quadrature	$E = \begin{pmatrix} \mathrm{Re}\left(E_x\right) \\ \mathrm{Im}\left(E_x\right) \\ \mathrm{Re}\left(E_y\right) \\ \mathrm{Im}\left(E_y\right) \end{pmatrix}$

$$\rho_1 = \begin{pmatrix} 0 & -1 & 0 & 0 \\ 1 & 0 & 0 & 0 \\ 0 & 0 & 0 & 1 \\ 0 & 0 & -1 & 0 \end{pmatrix} \quad \rho_2 = \begin{pmatrix} 0 & 0 & 0 & -1 \\ 0 & 0 & 1 & 0 \\ 0 & -1 & 0 & 0 \\ 1 & 0 & 0 & 0 \end{pmatrix} \quad \rho_3 = \begin{pmatrix} 0 & 0 & 1 & 0 \\ 0 & 0 & 0 & 1 \\ -1 & 0 & 0 & 0 \\ 0 & -1 & 0 & 0 \end{pmatrix}$$

$$(3.5)$$

$$\rho = \left(\rho_1, \rho_2, \rho_3\right) \tag{3.6}$$

$$\lambda_1 = \begin{pmatrix} 0 & 1 & 0 & 0 \\ -1 & 0 & 0 & 0 \\ 0 & 0 & 0 & 1 \\ 0 & 0 & -1 & 0 \end{pmatrix} \quad \lambda_2 = \begin{pmatrix} 0 & 0 & 0 & -1 \\ 0 & 0 & -1 & 0 \\ 0 & 1 & 0 & 0 \\ 1 & 0 & 0 & 0 \end{pmatrix} \quad \lambda_3 = \begin{pmatrix} 0 & 0 & 1 & 0 \\ 0 & 0 & 0 & -1 \\ -1 & 0 & 0 & 0 \\ 0 & 1 & 0 & 0 \end{pmatrix}$$

$$(3.7)$$

$$\lambda = \left(\lambda_1, \lambda_2, \lambda_3\right) \tag{3.8}$$

Any rotation R can be expressed in parametric form with respect to Equation (3.9):

$$R\left(\alpha, \beta\right) = e^{-\alpha \cdot \rho - \beta \cdot \lambda} \tag{3.9}$$

where $\boldsymbol{\alpha} = (\alpha_1, \alpha_2, \alpha_3)$ and $\boldsymbol{\beta} = (\beta_1, \beta_2, \beta_3)$ are the parameters.

For $\beta = 0$ the only contribution is given by ρ and we speak of right-iso-clinic rotations, that are two counter-directed simple rotations. Conversely, for $\rho = 0$ the only contribution is given by β and we speak of left-isoclinic rotations, that are two co-directed simple rotations.

Taking into account the properties of the exponential function, it is also possible to decompose the rotation matrix R into the two aforementioned contributions.

$$R(\alpha,\beta) = e^{-\alpha \cdot \rho - \beta \cdot \lambda} = R_R R_L = R_L R_R \qquad (3.10)$$

where

$$R_R(\alpha) = e^{-\alpha \cdot \rho} \qquad (3.11)$$

$$R_L(\beta) = e^{-\beta \cdot \lambda} \qquad (3.12)$$

As schematized in Table 3.2, the first four generating rotations (Equations [3.5] and [3.7]) are physically possible during the propagation of the electromagnetic field and correspond to the well-known phenomena of birefringence and phase shift, while the last two rotations are not physically possible during propagation.

This new formalism is very useful as it allows a physical phenomenon to be better represented mathematically which evidently, as shown since 1991, has a four-dimensional nature. Furthermore, the four degrees of freedom limit explains why the increase in the dimensionality of multilevel modulation schemes produces benefits only up to the fourth dimension and there are no further advantages in the higher dimensions.

The year after Arend et al. proposed satellite communication systems based on four-dimensional signals [19]. The authors started from the observation that satellite communication channel is similar to fiber. For example, satellite communication antennas are very direct and show a good polarization discrimination. Therefore, it is possible to use four-dimensional communication schemes on two orthogonal polarizations. Furthermore, the authors observed that from an engineering and practical point of view, the use of two orthogonal polarizations has advantages that add up to those of

Table 3.2 Four-dimensional rotations

Rotation class	Rotation matrix	Physical phenomenon	Time scale
Right-isoclinic	$e^{-\alpha_1 \rho_1}$	Linear birefringence	ms
	$e^{-\alpha_2 \rho_2}$	Linear birefringence	ms
	$e^{-\alpha_3 \rho_3}$	Circular birefringence	ms
Left-isoclinic	$e^{-\beta_1 \lambda_1}$	Phase shift	μs
	$e^{-\beta_2 \lambda_2}$	−	−
	$e^{-\beta_3 \lambda_3}$	−	−

the spectral efficiency already indicated in the works previously described. For example, the possibility of using the same local oscillator for both polarizations which therefore have the same phase, frequency and phase noise. Another important aspect is the possibility of carrying out a cross-polarization compensation for the extraction of the frequency, phase or clock at the receiver. The authors concluded that the use of four-dimensional communication schemes bring many small advantages which, even if they are minimal observed individually, overall offer an advantage over traditional communication systems which justifies their development and use.

3.3 PHASE AND POLARIZATION MULTIDIMENSIONAL MODULATION

In 2017 and 2018 Perrone et al. proposed new shift keying in optical fibers based on combined phase and polarization schemes [20, 21]. A description of [20] is presented in Section 2.5. The main idea behind these works is the observation that the four-dimensional space of information associated with the electromagnetic field can be decomposed into two three-dimensional spaces with a precise and easily measurable physical meaning: the polarization space and the phase space. The polarization space and its representation with the Poincaré sphere have already been discussed in Chapter 2. The phase space and the phase sphere have a construction similar to that of polarization and have been described by Born et al. [22] and used by Dennis [23] to study polarization singularities. In particular, the phase parameters that outline the phase sphere are:

$$
\begin{aligned}
T_0 &= S_0 = a^2 + b^2 + c^2 + d^2, \\
T_1 &= a^2 - b^2 + c^2 - d^2, \\
T_2 &= 2 \times (a \cdot b + c \cdot d), \\
T_3 &= S_3 = 2 \times (a \cdot d - b \cdot c).
\end{aligned}
\tag{3.13}
$$

Polarization and Phase parameters have a common term ($S_3 = T_3$). It is possible to combine polarization and phase points in order to obtain complex constellations. In [21] the authors propose two different constellations, the double tetrahedron and the double cube, illustrated respectively in Figures 3.6 and 3.7 and defined in Tables 3.3 and 3.4.

The transmitter for the transmission system in question can be the one already indicated in Figure 3.2, which allows the choice of the desired values for (a, b, c, d). The receiver, on the other hand, can be a simplified version of the one in 4-Quadrature receiver, at least as regards the processing of data in the baseband and the decision (Figure 3.8). In fact, once the values (a, b, c, d) have been identified, they can be combined to form the Stokes and Phase parameters. The former can be identified by appropriate tracking of

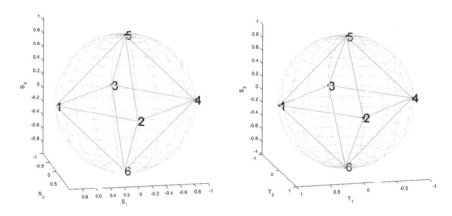

Figure 3.6 Double tetrahedron constellation on Poincaré and Phase spheres.

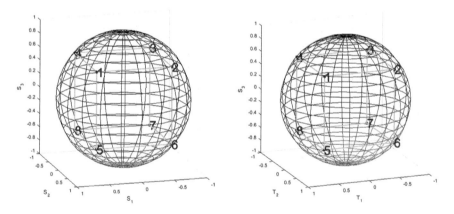

Figure 3.7 Double cube constellation on Poincaré and Phase spheres.

polarization variations, the others by tracking phase variations, both according to techniques already known in literature.

This communication system does not show a significant increase in performance compared to conventional modulation techniques or compared to the 4 Quadrature system. The merit of this work is instead that of showing that the four-dimensional transmission system 4-Quadrature Signaling, apparently far from the known optical-physical phenomena, can actually be decomposed into these systems. In the specific case described, the receiver can reconstruct a signal composed of phase and polarization shift keying or simply a signal modulated only in phase or only in polarization or even work with each of the three possible modulations at different times. All this happens with a single transmitter, the one described in Figure 3.2, which demonstrates its versatility in building signals by controlling all the possible parameters that characterize the electromagnetic field.

Table 3.3 Double tetrahedron constellation points

Combination	θ	φ_x	φ_y	S_1	S_2	S_3	T_1	T_2	T_3	x_1	x_2	x_3	x_4
4S-1T	0	0	0	-1	0	0	-1	0	0	0	0	-1	0
4S-2T	0	0	45	-1	0	0	0	-1	0	0	0	0,707	0,707
4S-4T	0	0	90	-1	0	0	-1	0	0	0	0	0	-1
4S-3T	0	0	135	-1	0	0	0	-1	0	0	0	-0,707	0,707
1S-1T	90	0	0	-1	0	0	-1	0	0	-1	0,707	0	0
1S-2T	90	45	0	-1	0	0	0	-1	0	0,707	-1	0	0
1S-4T	90	90	0	-1	0	0	-1	0	0	0	0,707	0	0
1S-3T	90	135	0	-1	0	0	0	-1	0	-0,707	0,5	0	0
2S-2T	45	45	45	0	-1	0	0	-1	0	0,5	0,707	0,5	0,5
2S-4T	45	90	90	0	-1	0	-1	0	0	0	0,5	0	0,707
2S-3T	45	135	135	0	-1	0	0	-1	0	-0,5	0	-0,5	0,5
2S-1T	45	180	180	0	-1	0	-1	0	0	-0,707	0,5	-0,707	0
3S-3T	-45	-45	-45	0	-1	0	0	-1	0	-0,5	0,707	0,5	-0,5
3S-4T	-45	-90	-90	0	-1	0	-1	0	0	0	0,5	0	-0,707
3S-2T	-45	-135	-135	0	-1	0	0	-1	0	0,5	0	-0,5	-0,5
3S-1T	-45	-180	-180	0	-1	0	-1	0	0	0,707	0	-0,707	0
5S-5T	45	0	90	0	0	-1	0	0	-1	0,707	0	0	0,707
6S-6T	-45	0	90	0	0	-1	0	0	-1	-0,707	0	0	0,707

Table 3.4 Double cube constellation points

Comb.	θ	φ_x	φ_y	S_1	S_2	S_3	T_1	T_2	T_3	x_1	x_2	x_3	x_4
1S-2T	62,632	60	105	0,577	0,577	0,577	-0,577	0,577	0,577	0,444	0,769	-0,119	0,444
1S-3T	62,632	105	150	0,577	0,577	0,577	-0,577	-0,577	0,577	-0,230	0,858	-0,398	0,230
1S-4T	62,632	150	195	0,577	0,577	0,577	0,577	-0,577	0,577	-0,769	0,444	-0,444	-0,119
1S-1T	62,632	195	240	0,577	0,577	0,577	0,577	0,577	0,577	-0,858	-0,230	-0,230	-0,398
4S-2T	62,632	75	210	0,577	-0,577	0,577	-0,577	0,577	0,577	0,230	0,858	-0,398	-0,230
4S-3T	62,632	120	255	0,577	-0,577	0,577	-0,577	-0,577	0,577	-0,444	0,769	-0,119	-0,444
4S-4T	62,632	165	300	0,577	-0,577	0,577	0,577	-0,577	0,577	-0,858	0,230	0,230	-0,398
4S-1T	62,632	30	165	0,577	-0,577	0,577	0,577	0,577	0,577	0,769	0,444	-0,444	0,119
2S-2T	27,368	30	75	0,577	0,577	0,577	-0,577	0,577	0,577	0,398	0,230	0,230	0,858
2S-3T	27,368	75	120	-0,577	0,577	0,577	-0,577	-0,577	0,577	0,119	0,444	-0,444	0,769
2S-4T	27,368	120	165	-0,577	0,577	0,577	0,577	-0,577	0,577	-0,230	0,398	-0,858	0,230
2S-1T	27,368	165	210	-0,577	0,577	0,577	0,577	0,577	0,577	-0,444	0,119	-0,769	-0,444
3S-2T	27,368	105	240	-0,577	-0,577	0,577	-0,577	0,577	0,577	-0,119	0,444	-0,444	-0,769
3S-3T	27,368	150	285	-0,577	-0,577	0,577	-0,577	-0,577	0,577	-0,398	0,230	0,230	-0,858
3S-4T	27,368	195	330	-0,577	-0,577	0,577	0,577	-0,577	0,577	-0,444	-0,119	0,769	-0,444
3S-1T	27,368	60	195	-0,577	-0,577	0,577	0,577	0,577	0,577	0,230	0,398	-0,858	-0,230
5S-6T	-62,632	75	210	0,577	0,577	-0,577	-0,577	0,577	-0,577	-0,230	-0,858	-0,398	-0,230
5S-7T	-62,632	120	255	0,577	0,577	-0,577	-0,577	-0,577	-0,577	0,444	-0,769	-0,119	-0,444
5S-8T	-62,632	165	300	0,577	0,577	-0,577	0,577	-0,577	-0,577	0,858	-0,230	0,230	-0,398
5S-5T	-62,632	30	165	0,577	0,577	-0,577	0,577	0,577	-0,577	-0,769	-0,444	-0,444	0,119

(Continued)

Table 3.4 (Continued) Double cube constellation points

Comb.	θ	φ_x	φ_y	S_1	S_2	S_3	T_1	T_2	T_3	x_1	x_2	x_3	x_4
8S-6T	−62,632	60	105	0,577	−0,577	−0,577	−0,577	0,577	−0,577	−0,444	−0,769	−0,119	0,444
8S-7T	−62,632	105	150	0,577	−0,577	−0,577	−0,577	−0,577	−0,577	0,230	−0,858	−0,398	0,230
8S-8T	−62,632	150	195	0,577	−0,577	−0,577	0,577	−0,577	−0,577	0,769	−0,444	−0,444	−0,119
8S-5T	−62,632	195	240	0,577	−0,577	−0,577	0,577	0,577	−0,577	0,858	0,230	−0,230	−0,398
6S-6T	−27,368	105	240	−0,577	0,577	−0,577	−0,577	0,577	−0,577	0,119	−0,444	−0,444	−0,769
6S-7T	−27,368	150	285	−0,577	0,577	−0,577	−0,577	−0,577	−0,577	0,398	−0,230	0,230	−0,858
6S-8T	−27,368	195	330	−0,577	0,577	−0,577	0,577	−0,577	−0,577	0,444	0,119	0,769	−0,444
6S-5T	−27,368	60	195	−0,577	0,577	−0,577	0,577	0,577	−0,577	−0,230	−0,398	−0,858	−0,230
7S-6T	−27,368	30	75	−0,577	−0,577	−0,577	−0,577	0,577	−0,577	−0,398	−0,230	0,230	0,858
7S-7T	−27,368	75	120	−0,577	−0,577	−0,577	−0,577	−0,577	−0,577	−0,119	−0,444	−0,444	0,769
7S-8T	−27,368	120	165	−0,577	−0,577	−0,577	0,577	−0,577	−0,577	0,230	−0,398	−0,858	0,230
7S-5T	−27,368	165	210	−0,577	−0,577	−0,577	0,577	0,577	−0,577	0,444	−0,119	−0,769	−0,444

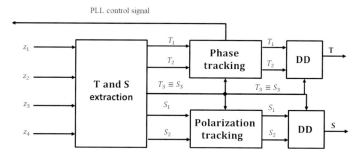

Figure 3.8 Baseband section for a Phase and Polarization receiver. DD: Decision Device.

3.4 REPRESENTATIONS OF OPTICAL FIBER COMMUNICATIONS IN FOUR- AND THREE-DIMENSIONAL SPACES

In telecommunications, some quantities of a physical phenomenon are used to transmit information. Amplitude, frequency, phase, quadrature, polarization or some combinations thereof are commonly used electromagnetic signal features. Moreover, it is possible to use multiple domains at the same time, exploiting the partial or complete independence between them, in order to maximize the electromagnetic field capacity to carry information. This generates hybrid modulations, such as phase-quadrature, polarization-quadrature-amplitude and so on. Therefore, the transmission system proposed in [20] and [21] is not an absolute novelty but only a new hybrid transmission scheme which is added to the others already present in the literature. As already mentioned, the importance of the latter, however, lies in the strong analogies between the equations that represent Stokes' parameters and the phase parameters that define their respective spheres. These analogies suggest to analyze the mathematical formulas of these representations, to construct others by analogy and extension and to relate them to each other and with the four-dimensional representation already described, the 4-Quadrature Signaling. This study was done in 2018–19 by Rutigliano et al. [24, 25].

We rewrite the Stokes and phase parameters as a function of the four components of the electromagnetic field identified by Equation (3.2):

$$
\begin{aligned}
S_0 &= a^2 + b^2 + c^2 + d^2 \\
S_1 &= a^2 + b^2 - c^2 - d^2 \\
S_2 &= 2 \times (a \cdot c + b \cdot d) \\
S_3 &= 2 \times (a \cdot d - b \cdot c)
\end{aligned}
\tag{3.14}
$$

$$T_0 = a^2 + b^2 + c^2 + d^2$$
$$T_1 = a^2 - b^2 + c^2 - d^2$$
$$T_2 = 2 \times (a \cdot b + c \cdot d)$$
$$T_3 = 2 \times (a \cdot d - b \cdot c).$$

(3.15)

The observation of the quadratic and bilinear terms present in the above equations allows us to identify a symmetry in their construction. From a geometric point of view we can consider these equations as transformations that project a four-dimensional space (a, b, c, d) into two different three-dimensional spaces, respectively (S_1, S_2, S_3) and (T_1, T_2, T_3).

Also, we can add that the four-dimensional space (a, b, c, d) can be projected on four three-dimensional spaces (the four possible triples with a, b, c and d) and six two-dimensional spaces (the six possible couples with a, b, c and d). Tables 3.5 and 3.6 show the bilinear terms necessary to fully represent the S and T parameters.

The quadratic terms (a^2, b^2, c^2 and d^2) can be represented on one of the planes that contain the corresponding letter.

Observing Tables 3.5 and 3.6 we can state that:

1. Each representation uses four of the six available planes.
2. Each plane presents two adjacent planes.
3. Each axis appears two times.

Such symmetries and analogies are the direct consequence of the fact that the two representations are transformed projections of the same phenomenon.

Table 3.5 Stokes' parameters

	ab	ac	ad	bc	bd	cd
S_0		X			X	
S_1		X			X	
S_2		X			X	
S_3			X	X		

Table 3.6 Phase parameters

	ab	ac	ad	bc	bd	cd
T_0	X					X
T_1	X					X
T_2	X					X
T_3			X	X		

Table 3.7 R parameters

	ab	ac	ad	bc	bd	cd
R_0	X					X
R_1	X					X
R_2	X					X
R_3		X			X	

Table 3.8 P parameters

	ab	ac	ad	bc	bd	cd
P_0	X					X
P_1	X					X
P_2		X			X	
P_3	X					X

Using the same properties, and keeping the considerations so far made, we can build a third and a fourth representation of the electromagnetic field whose terms, respectively called R and P parameters, are represented in Tables 3.7 and 3.8.

The representation according to the parameters R is defined as follows:

$$
\begin{aligned}
R_0 &= a^2 + b^2 + c^2 + d^2 \\
R_1 &= -a^2 + b^2 + c^2 - d^2 \\
R_2 &= 2 \times (a \cdot b + c \cdot d) \\
R_3 &= 2 \times (a \cdot c - b \cdot d).
\end{aligned}
\tag{3.16}
$$

The representation according to the parameters P is defined as follows:

$$
\begin{aligned}
P_0 &= a^2 + b^2 + c^2 + d^2 \\
P_1 &= -a^2 + b^2 + c^2 - d^2 \\
P_2 &= 2 \times (a \cdot c + b \cdot d) \\
P_3 &= 2 \times (a \cdot b - c \cdot d).
\end{aligned}
\tag{3.17}
$$

Such representations, formally analogous to the other two, respect the power constraint. That is, the following relations apply:

$$
\begin{aligned}
S_0^2 &= S_1^2 + S_2^2 + S_3^2 \\
T_0^2 &= T_1^2 + T_2^2 + T_3^2 \\
R_0^2 &= R_1^2 + R_2^2 + R_3^2 \\
P_0^2 &= P_1^2 + P_2^2 + P_3^2.
\end{aligned}
\tag{3.18}
$$

So, even in the spaces (R_1, R_2, R_3) and (P_1, P_2, P_3) the values of power of the electromagnetic field, respectively R_0^2 and P_0^2, lie on the surfaces of the spheres of radius, R_0 and P_0, respectively. It is worth stressing that, in order to comply with (3.18), the bilinear terms present in R_2 and R_3 must be composed of couples of letters whose squares have opposite sign in R_1. This consideration also applies to parameters S, T and P.

Finally, we note that in the construction of the representations, given the presence of quadratic terms and products, it is possible to obtain another 12 representations that differ in some signs. Indeed, for our purposes, these representations can be considered linearly dependent on the ones so far described. Changing some signs in bilinear terms we get other four representations that we indicate with the parameters S2, T2, R2, P2:

$$
\begin{aligned}
S2_0 &= a^2 + b^2 + c^2 + d^2 \\
S2_1 &= a^2 + b^2 - c^2 - d^2 \\
S2_2 &= 2 \times (a \cdot c - b \cdot d) \\
S2_3 &= 2 \times (a \cdot d + b \cdot c)
\end{aligned}
\tag{3.19}
$$

$$
\begin{aligned}
T2_0 &= a^2 + b^2 + c^2 + d^2 \\
T2_1 &= a^2 - b^2 + c^2 - d^2 \\
T2_2 &= 2 \times (a \cdot b - c \cdot d) \\
T2_3 &= 2 \times (a \cdot d + b \cdot c)
\end{aligned}
\tag{3.20}
$$

$$
\begin{aligned}
R2_0 &= a^2 + b^2 + c^2 + d^2 \\
R2_1 &= -a^2 + b^2 + c^2 - d^2 \\
R2_2 &= 2 \times (a \cdot b - c \cdot d) \\
R2_3 &= 2 \times (a \cdot c + b \cdot d)
\end{aligned}
\tag{3.21}
$$

$$
\begin{aligned}
P2_0 &= a^2 + b^2 + c^2 + d^2 \\
P2_1 &= -a^2 + b^2 + c^2 - d^2 \\
P2_2 &= 2 \times (a \cdot c - b \cdot d) \\
P2_3 &= 2 \times (a \cdot b + c \cdot d).
\end{aligned}
\tag{3.22}
$$

Starting from (3.14), (3.15), (3.16), (3.17), (3.19), (3.20), (3.21), (3.22) and reversing the signs of the quadratic terms of the parameters with subscript 1 we obtain the other eight representations that we indicate with S3, T3, R3, P3, S4, T4, R4 and P4:

$$
\begin{aligned}
S3_0 &= a^2 + b^2 + c^2 + d^2 \\
S3_1 &= -a^2 - b^2 + c^2 + d^2 \\
S3_2 &= 2 \times (a \cdot c + b \cdot d) \\
S3_3 &= 2 \times (a \cdot d - b \cdot c)
\end{aligned}
\tag{3.23}
$$

$$T3_0 = a^2 + b^2 + c^2 + d^2$$
$$T3_1 = -a^2 + b^2 - c^2 + d^2$$
$$T3_2 = 2 \times (a \cdot b + c \cdot d)$$
$$T3_3 = 2 \times (a \cdot d - b \cdot c)$$

(3.24)

$$R3_0 = a^2 + b^2 + c^2 + d^2$$
$$R3_1 = +a^2 - b^2 - c^2 + d^2$$
$$R3_2 = 2 \times (a \cdot b + c \cdot d)$$
$$R3_3 = 2 \times (a \cdot c - b \cdot d)$$

(3.25)

$$P3_0 = a^2 + b^2 + c^2 + d^2$$
$$P3_1 = +a^2 - b^2 - c^2 + d^2$$
$$P3_2 = 2 \times (a \cdot c + b \cdot d)$$
$$P3_3 = 2 \times (a \cdot b - c \cdot d)$$

(3.26)

$$S4_0 = a^2 + b^2 + c^2 + d^2$$
$$S4_1 = -a^2 - b^2 + c^2 + d^2$$
$$S4_2 = 2 \times (a \cdot c - b \cdot d)$$
$$S4_3 = 2 \times (a \cdot d + b \cdot c)$$

(3.27)

$$T4_0 = a^2 + b^2 + c^2 + d^2$$
$$T4_1 = -a^2 + b^2 - c^2 + d^2$$
$$T4_2 = 2 \times (a \cdot b - c \cdot d)$$
$$T4_3 = 2 \times (a \cdot d + b \cdot c)$$

(3.28)

$$R4_0 = a^2 + b^2 + c^2 + d^2$$
$$R4_1 = -a^2 + b^2 + c^2 - d^2$$
$$R4_2 = 2 \times (a \cdot b - c \cdot d)$$
$$R4_3 = 2 \times (a \cdot c + b \cdot d)$$

(3.29)

$$P4_0 = a^2 + b^2 + c^2 + d^2$$
$$P4_1 = +a^2 - b^2 - c^2 + d^2$$
$$P4_2 = 2 \times (a \cdot c - b \cdot d)$$
$$P4_3 = 2 \times (a \cdot b + c \cdot d).$$

(3.30)

It is immediate to verify that all representations respect the constraints on power in analogy to Equation (3.18). That is, the following relations apply:

$$S2_0^2 = S2_1^2 + S2_2^2 + S2_3^2$$
$$T2_0^2 = T2_1^2 + T2_2^2 + T2_3^2$$
$$R2_0^2 = R2_1^2 + R2_2^2 + R2_3^2$$
$$P2_0^2 = P2_1^2 + P2_2^2 + P2_3^2$$
$$S3_0^2 = S3_1^2 + S3_2^2 + S3_3^2$$
$$T3_0^2 = T3_1^2 + T3_2^2 + T3_3^2 \tag{3.31}$$
$$R3_0^2 = R3_1^2 + R3_2^2 + R3_3^2$$
$$P3_0^2 = P3_1^2 + P3_2^2 + P3_3^2$$
$$S4_0^2 = S4_1^2 + S4_2^2 + S4_3^2$$
$$T4_0^2 = T4_1^2 + T4_2^2 + T4_3^2$$
$$R4_0^2 = R4_1^2 + R4_2^2 + R4_3^2$$
$$P4_0^2 = P4_1^2 + P4_2^2 + P4_3^2.$$

We started from the four-dimensional hyperspace defined by the phase and quadrature components of the electromagnetic field components E_x and E_y. We showed that this hyperspace, used in the 4-Quadrature Signaling, could be projected into four transformed three-dimensional spaces, defined by Equations (3.14)–(3.17). In these four transformed spaces different characteristics of the same phenomenon, the electromagnetic field, are highlighted. The physical significance of the S and T parameters is already known and relates respectively to the polarization and the phase of the electromagnetic field. The physical meaning of the new R and P parameters can be investigated by recalling the definition of a, b, c and d (E_{xr}, E_{xi}, E_{yr}, E_{yi}).

In the following, we indicated with v the module of vector **v**, with **v*** the conjugate complex of **v** and with **v'** the vector, or complex number, with real and imaginary part inverted compared to **v**. Moreover, we define:

$$x = [a\,b] = [E_{xr}\ E_{xi}]$$
$$y = [c\,d] = [E_{yr}\ E_{yi}]$$
$$p = [a\,c] = [E_{xr}\ E_{yr}] \tag{3.32}$$
$$q = [b\,d] = [E_{xi}\ E_{yi}].$$

Taking into account the above definitions we can rewrite (3.14)–(3.17) as follows:

$$S_o = a^2 + b^2 + c^2 + d^2$$
$$S_1 = a^2 + b^2 - c^2 - d^2 = x^2 - y^2$$
$$S_2 = 2 \times (a \cdot c + b \cdot d) = 2 \times (x \cdot y) \tag{3.33}$$
$$S_3 = 2 \times (a \cdot d - b \cdot c) = 2 \times (x \cdot y^{*'})$$

$$T_o = a^2 + b^2 + c^2 + d^2$$
$$T_1 = a^2 - b^2 + c^2 - d^2 = p^2 - q^2$$
$$T_2 = 2 \times (a \cdot b + c \cdot d) = 2 \times (p \cdot q) \qquad\qquad (3.34)$$
$$T_3 = 2 \times (a \cdot d - b \cdot c) = 2 \times (p \cdot q^*)$$

$$R_o = a^2 + b^2 + c^2 + d^2$$
$$R_1 = -a^2 + b^2 + c^2 - d^2 = -x \cdot x^* + y \cdot y^*$$
$$R_2 = 2 \times (a \cdot b + c \cdot d) = 2 \times (p \cdot q) \qquad\qquad (3.35)$$
$$R_3 = 2 \times (a \cdot c - b \cdot d) = 2 \times (x \cdot y')$$

$$P_o = a^2 + b^2 + c^2 + d^2$$
$$P_1 = -a^2 + b^2 + c^2 - d^2 = -x \cdot x^* + y \cdot y^*$$
$$P_2 = 2 \times (a \cdot c + b \cdot d) = 2 \times (x \cdot y) \qquad\qquad (3.36)$$
$$P_3 = 2 \times (a \cdot b - c \cdot d) = 2 \times (p \cdot q').$$

We therefore have confirmation that the Stokes' parameters explicitly contain information about the polarization of the electromagnetic field. Likewise, the phase parameters explicitly contain information regarding the phase of the electromagnetic field. The R and P parameters, on the other hand, are hybrids as they contain information that can be immediately traced back to both the phase and the polarization of the electromagnetic. For the sake of brevity we do not treat the other 12 representations also because many parameters are the same or simply of opposite sign. Indeed it is easy to verify that the following relationships apply:

$$S_1 = S2_1 = -S3_1 = S4_1$$
$$S_2 = P_2 = S3_2 = R2_3 = R4_3 = P3_2$$
$$S_3 = T_3 = S3_3 = T3_3$$
$$S2_3 = S4_3 = T2_3 = T4_3$$
$$T_1 = T2_1 = -T3_1 = -T4_1$$
$$T_2 = R_2 = T3_2 = R3_2 = P2_3 = P4_3 \qquad\qquad (3.37)$$
$$R_1 = R2_1 = -R3_1 = -R4_1$$
$$R_3 = S2_2 = S4_2 = R3_3 = P2_2 = P4_2$$
$$P_1 = P2_1 = -P3_1 = -P4_1$$
$$P_3 = P3_3 = T2_2 = T4_2 = R2_2 = R4_2.$$

In Figure 3.9 the different representations of a four-dimensional antipodal coding are shown. The two symbols 0 ($+\alpha, +\alpha, +\alpha, +\alpha$) and 1 ($-\alpha, -\alpha, -\alpha, -\alpha$)

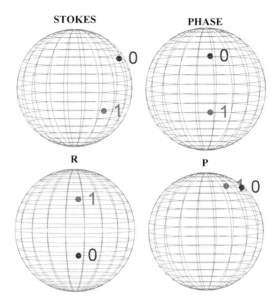

Figure 3.9 Antipodal four-dimensional symbols projected into the four three-dimensional spaces.

of the four-dimensional Cartesian space are transformed into the four three-dimensional spaces of Stokes, Phases, R and P and represented on the respective spheres.

We can observe that in all three-dimensional representations the symbols are not antipodal. This is due to the fact that the transformations defined by (3.14)–(3.17) are neither linear nor isomorphic, that is they do not preserve the morphological characteristics of the constellation used in the four-dimensional space. It is also easy to verify that the distances between the two symbols are lower in the three-dimensional spaces than those in the four-dimensional space [26].

In Figure 3.10 the four-dimensional hypercube, the tesseract, is represented using the four three-dimensional representations. In particular, these transformed projections refer to a tesseract rotated (90,0,0) degrees with respect to the axis of the four-dimensional space: in this case the 16 symbols, the vertexes of the tesseract, assume eight different positions and describe circumferences in all the spaces. Note that with more symbols, they overlap in pairs due to the loss of a space dimension (from 4 to 3) and it means the loss of half of the transmission capacity. However, we can graphically observe that the positions of the symbols maintain their orthogonality in the different spaces and also the constraints on the common axes. For example, on the Stokes and phase spheres we can verify that the symbols are positioned at the same height. This confirms that the values of S_3 and P_3 coincide as expected from 3.13 and 3.37.

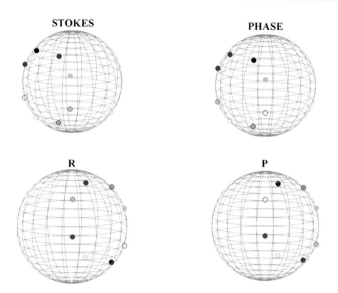

Figure 3.10 Tesseract four-dimensional symbols projected into the four three-dimensional spaces.

To better understand the relationships between the different representations, in the following we write the parameters in matrix form. For example, for S_1 we can write:

$$S_1 = [a\,b\,c\,d] \times \begin{bmatrix} 1 & 0 & 0 & 0 \\ 0 & 1 & 0 & 0 \\ 0 & 0 & -1 & 0 \\ 0 & 0 & 0 & -1 \end{bmatrix} \times \begin{bmatrix} a \\ b \\ c \\ d \end{bmatrix} = a^2 + b^2 - c^2 - d^2. \qquad (3.38)$$

The matrices corresponding to the S, T, R and P parameters, showed in Table 3.9, are symmetric, unitary, rank 4 and are linearly independent, with four exceptions (four constraints, excluding power):

$S_0 = T_0 = R_0 = P_0$

$S_2 = P_2$

$S_3 = T_3$ $\qquad\qquad\qquad\qquad\qquad\qquad\qquad (3.39)$

$T_2 = R_2$

$R_1 = P_1.$

This confirms that the four three-dimensional spaces are not independent but each has one axis in common with other two spaces, as it is shown in Figure 3.11 for the first four representations and in Figure 3.12 for the second four representations. Analog relationships apply for the quartets S3, T3, R3, P3 and S4, T4, R4, P4 (Table 3.10).

Table 3.9 Matrix form of the first four representations

	0	1	2	3
S	$\begin{bmatrix} I & & & \\ & I & & \\ & & I & \\ & & & I \end{bmatrix}$	$\begin{bmatrix} I & & & \\ & I & & \\ & & -I & \\ & & & -I \end{bmatrix}$	$\begin{bmatrix} & & I & \\ & & & I \\ I & & & \\ & I & & \end{bmatrix}$	$\begin{bmatrix} & & & I \\ & & -I & \\ & -I & & \\ I & & & \end{bmatrix}$
T	$\begin{bmatrix} I & & & \\ & I & & \\ & & I & \\ & & & I \end{bmatrix}$	$\begin{bmatrix} I & & & \\ & -I & & \\ & & I & \\ & & & -I \end{bmatrix}$	$\begin{bmatrix} & & I & \\ I & & & \\ & & & I \\ & I & & \end{bmatrix}$	$\begin{bmatrix} & & & I \\ & & -I & \\ -I & & & \\ & I & & \end{bmatrix}$
R	$\begin{bmatrix} I & & & \\ & I & & \\ & & I & \\ & & & I \end{bmatrix}$	$\begin{bmatrix} -I & & & \\ & I & & \\ & & I & \\ & & & -I \end{bmatrix}$	$\begin{bmatrix} & & I & \\ & & & I \\ I & & & \\ & I & & \end{bmatrix}$	$\begin{bmatrix} & & I & \\ & & & -I \\ I & & & \\ & -I & & \end{bmatrix}$
P	$\begin{bmatrix} I & & & \\ & I & & \\ & & I & \\ & & & I \end{bmatrix}$	$\begin{bmatrix} -I & & & \\ & I & & \\ & & I & \\ & & & -I \end{bmatrix}$	$\begin{bmatrix} & & I & \\ & & & I \\ I & & & \\ & I & & \end{bmatrix}$	$\begin{bmatrix} & & I & \\ & & & -I \\ I & & & \\ & -I & & \end{bmatrix}$

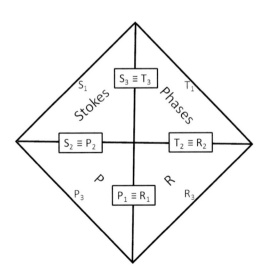

Figure 3.11 Relationships between the axis of the first four three-dimensional representations.

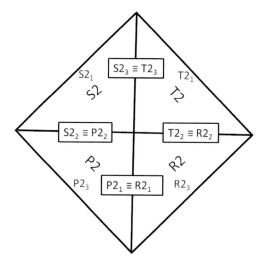

Figure 3.12 Relationships between the axis of the second four three-dimensional representations.

Table 3.10 Matrix form of the other representations

	0	1	2	3
S2	$\begin{bmatrix} 1 & & & \\ & 1 & & \\ & & 1 & \\ & & & 1 \end{bmatrix}$	$\begin{bmatrix} 1 & & & \\ & 1 & & \\ & & -1 & \\ & & & -1 \end{bmatrix}$	$\begin{bmatrix} & & 1 & \\ 1 & & & \\ & -1 & & \\ & & & -1 \end{bmatrix}$	$\begin{bmatrix} & & & 1 \\ & & 1 & \\ & 1 & & \\ 1 & & & \end{bmatrix}$
T2	$\begin{bmatrix} 1 & & & \\ & 1 & & \\ & & 1 & \\ & & & 1 \end{bmatrix}$	$\begin{bmatrix} 1 & & & \\ & -1 & & \\ & & 1 & \\ & & & -1 \end{bmatrix}$	$\begin{bmatrix} & & 1 & \\ 1 & & & \\ & & & -1 \\ & -1 & & \end{bmatrix}$	$\begin{bmatrix} & & & 1 \\ & & 1 & \\ & 1 & & \\ 1 & & & \end{bmatrix}$
R2	$\begin{bmatrix} 1 & & & \\ & 1 & & \\ & & 1 & \\ & & & 1 \end{bmatrix}$	$\begin{bmatrix} -1 & & & \\ & 1 & & \\ & & 1 & \\ & & & -1 \end{bmatrix}$	$\begin{bmatrix} & & 1 & \\ 1 & & & \\ & & & -1 \\ & -1 & & \end{bmatrix}$	$\begin{bmatrix} & & & 1 \\ & & 1 & \\ & 1 & & \\ 1 & & & \end{bmatrix}$
P2	$\begin{bmatrix} 1 & & & \\ & 1 & & \\ & & 1 & \\ & & & 1 \end{bmatrix}$	$\begin{bmatrix} -1 & & & \\ & 1 & & \\ & & 1 & \\ & & & -1 \end{bmatrix}$	$\begin{bmatrix} & & 1 & \\ 1 & & & \\ & -1 & & \\ & & & -1 \end{bmatrix}$	$\begin{bmatrix} & & & 1 \\ & 1 & & \\ & & 1 & \\ 1 & & & \end{bmatrix}$

(*Continued*)

Table 3.10 (Continued) Matrix form of the other representations

	0	1	2	3
S3	$\begin{bmatrix} 1&&&\\ &1&&\\ &&1&\\ &&&1 \end{bmatrix}$	$\begin{bmatrix} -1&&&\\ &-1&&\\ &&1&\\ &&&1 \end{bmatrix}$	$\begin{bmatrix} &&1&\\ &&&-1\\ 1&&&\\ &-1&& \end{bmatrix}$	$\begin{bmatrix} &&&1\\ &&1&\\ &1&&\\ 1&&& \end{bmatrix}$
T3	$\begin{bmatrix} 1&&&\\ &1&&\\ &&1&\\ &&&1 \end{bmatrix}$	$\begin{bmatrix} -1&&&\\ &1&&\\ &&-1&\\ &&&1 \end{bmatrix}$	$\begin{bmatrix} 1&&&\\ &&1&\\ &1&&\\ &&&1 \end{bmatrix}$	$\begin{bmatrix} &&&1\\ &&-1&\\ &-1&&\\ 1&&& \end{bmatrix}$
R3	$\begin{bmatrix} 1&&&\\ &1&&\\ &&1&\\ &&&1 \end{bmatrix}$	$\begin{bmatrix} 1&&&\\ &-1&&\\ &&-1&\\ &&&1 \end{bmatrix}$	$\begin{bmatrix} 1&&&\\ &&1&\\ &1&&\\ &&&1 \end{bmatrix}$	$\begin{bmatrix} &&1&\\ &&&-1\\ 1&&&\\ &-1&& \end{bmatrix}$
P3	$\begin{bmatrix} 1&&&\\ &1&&\\ &&1&\\ &&&1 \end{bmatrix}$	$\begin{bmatrix} 1&&&\\ &-1&&\\ &&-1&\\ &&&1 \end{bmatrix}$	$\begin{bmatrix} &&1&\\ &&1&\\ 1&&&\\ &-1&& \end{bmatrix}$	$\begin{bmatrix} &1&&\\ &&&-1\\ 1&&&\\ &&-1& \end{bmatrix}$
S4	$\begin{bmatrix} 1&&&\\ &1&&\\ &&1&\\ &&&1 \end{bmatrix}$	$\begin{bmatrix} -1&&&\\ &-1&&\\ &&1&\\ &&&1 \end{bmatrix}$	$\begin{bmatrix} &&1&\\ &&&-1\\ 1&&&\\ &-1&& \end{bmatrix}$	$\begin{bmatrix} &&&1\\ &&1&\\ &1&&\\ 1&&& \end{bmatrix}$
T4	$\begin{bmatrix} 1&&&\\ &1&&\\ &&1&\\ &&&1 \end{bmatrix}$	$\begin{bmatrix} -1&&&\\ &1&&\\ &&-1&\\ &&&1 \end{bmatrix}$	$\begin{bmatrix} 1&&&\\ &1&&\\ &&&-1\\ &&-1& \end{bmatrix}$	$\begin{bmatrix} &&&1\\ &&1&\\ &1&&\\ 1&&& \end{bmatrix}$
R4	$\begin{bmatrix} 1&&&\\ &1&&\\ &&1&\\ &&&1 \end{bmatrix}$	$\begin{bmatrix} 1&&&\\ &-1&&\\ &&-1&\\ &&&1 \end{bmatrix}$	$\begin{bmatrix} 1&&&\\ &1&&\\ &&&-1\\ &&-1& \end{bmatrix}$	$\begin{bmatrix} &&&1\\ 1&&&\\ &1&&\\ &&1& \end{bmatrix}$
P4	$\begin{bmatrix} 1&&&\\ &1&&\\ &&1&\\ &&&1 \end{bmatrix}$	$\begin{bmatrix} 1&&&\\ &-1&&\\ &&-1&\\ &&&1 \end{bmatrix}$	$\begin{bmatrix} &&1&\\ &&&-1\\ 1&&&\\ &-1&& \end{bmatrix}$	$\begin{bmatrix} &1&&\\ 1&&&\\ &&&1\\ &&1& \end{bmatrix}$

It is possible to verify that the degree of freedom of the system is four. In fact, keeping the power constant, the four spherical surfaces, each with two degrees of freedom (latitude and longitude angles), have a total of eight (2×4) degrees of freedom: these should be deducted from the number of constraints obtaining 8−4 = 4.

We have so far seen that a three-dimensional representation of the electromagnetic field, for example that of Stokes, does not fully exploit the four degrees of freedom available of the electromagnetic field. We have also seen how it is possible to combine two different representations, for example Stokes and phases [20, 21], to better exploit the capacity of the electromagnetic field. We now have 16 three-dimensional representations of the electromagnetic field that we can combine with each other but we have seen that they are also linearly dependent. So, it makes sense to ask to what extent it is possible to combine these representations or how many of them can be combined before the further combination does not bring an increase in the information carried by the electromagnetic field. To answer this question, let's think about the degrees of freedom offered by representations, taken individually or combined with each other.

One representation has two degrees of freedom. We can intuitively identify these two degrees of freedom as the latitude and longitude of the point identified on the surface of the sphere, for example that of Poincarè. In a more rigorous way we can observe that a representation, for example that of Stokes, has three parameters that can vary (S_1, S_2, S_3) but they are linked by a constraint, the one on the power (3.31) which subtracts a degree of freedom and causes the third parameter to be univocally determined once the first two have been set. Without losing generality, we consider the transmitted power to be constant, a condition corresponding to practical applications and therefore we do not consider the term S_0 to be variable.

With two representations the degrees of freedom become three. For example, if we use Stokes and Phase spaces [20, 21], we have six parameters (S_1, S_2, S_3, T_1, T_2, T_3) with two internal constraints in the spaces, given by (3.31) and one external, that is $S_3 = T_3$ (3.37). We can formally express this dependence:

$$S_1 = f_S\left(S_2, S_3 \equiv T_3\right) = \pm\sqrt{S_0^2 - S_2^2 - S_3^2}$$
$$T_1 = f_T\left(T_2, T_3 \equiv S_3\right). \tag{3.40}$$

The independent variables are three (S_2, S_3, T_2) and therefore the degrees of freedom are three.

With three representations, the number of degrees of freedom become four. For example, using spaces defined by S, T and R we have nine parameters, with three constraints internal to the spaces (3.31) and two external (3.37), $S_3 = T_3$ and $T_2 = R_2$. We can write:

$$S_1 = f_S\left(S_2, S_3 \equiv T_3\right)$$
$$T_1 = f_T\left(T_2, T_3 \equiv S_3\right)$$
$$R_1 = f_R\left(R_2 \equiv T_2, R_3\right). \tag{3.41}$$

The independent variables are four: S_2, S_3, T_2 and R_3. Such a transmission system could equal the performance of a 4-Quadrature Signaling transmission system in terms of degrees of freedom and information transport capacity.

Using more than three different spaces to represent information would be useless, since any fourth space would cause redundant information over the first three spaces. In fact, this result can also be deduced from the equivalence of Tables 3.7 and 3.8, which show how the representations based on the R and P parameters carry the same information content.

The fact that the combined use of more than three three-dimensional representations is not useful in increasing the amount of information transmitted does not mean that the 16 formally correct representations described so far cannot be exploited in order, for instance, to make transmission of information more secure or resilient.

3.5 APPLICATIONS

The results presented so far demonstrate the close link between the three- and four-dimensional representations of the electromagnetic field. From a physics point of view this fact is intuitive, because they are representations of the same physical phenomenon. From a mathematic point of view studying Tables 3.5 to 3.8 and (3.33), (3.34), (3.35), (3.36) we can observe that all the representations are made by ten fundamental elements of the second order (a^2, b^2, c^2, d^2 and ab, ac, ad, bc, bd, cd) and differ only for the presence of some elements or the sign with which these elements are present. This fact suggests the possibility of using a 4-Quadrature signaling transmission system [6] and of pre-elaborating and post-elaborating the values of the four values a, b, c, d in order to use one or more three-dimensional representation [20, 21].

The use of different representations in a transmission system, simultaneously or at different times, has the advantage of increasing transmission capacity compared to the current systems, as demonstrated in [20, 21], or in the possibility of choosing the representation affected by less noise, permanently or over time. For instance, phase noise has a minimal impact on a transmission system that exploits the polarization of the light and vice versa, as well as the two noises act on very different time scales, respectively of the order of microseconds and seconds.

In Figure 3.13 it is shown the effect of simulated phase noise on the first eight representations. We can observe that phase noise has negligible effect on Stokes and S2 representations and the same is valid for S3 and S4 parameters not shown for brevity. Phase noise has effects on phase based representation, that is Phase, T2, T3 and T4 but also on hybrid representations. Moreover, from the figure it is also possible to see the partial orthogonality of the different representations. In fact, in each representation a figure different in position and direction is described.

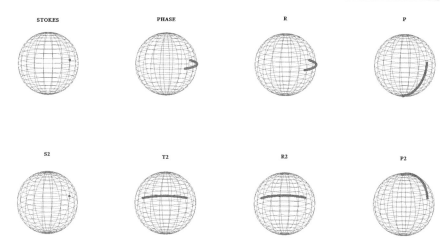

Figure 3.13 Effect of simulated phase noise on different electromagnetic-field representations.

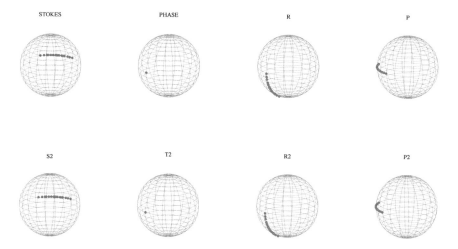

Figure 3.14 Effect of simulated circular birefringence polarization fluctuations on different electromagnetic-field representations.

In Figure 3.14 it is shown the effect of simulated circular birefringence polarization fluctuations on the first eight representations. This situation, already studied in [20], is dual with respect to that of Figure 3.13. Indeed, that perturbation has no effect on phase-based representations while it has effects on polarization based or hybrid ones.

The situation represented in Figure 3.15 is different: a non-circular polarization fluctuation is simulated but given by the sum of random horizontal and vertical linear polarization along the optical fiber, a situation that corresponds to the real case of the non-twisted optical fiber. The effect is

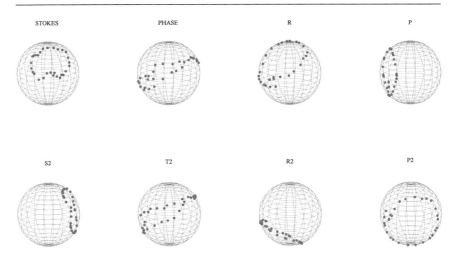

Figure 3.15 Effect of simulated linear horizontal and vertical birefringence polarization fluctuations on different electromagnetic-field representations.

certainly more chaotic and not easily separable in the different representations, although the orthogonality of the figures described on the different spheres remains valid. It is worth remembering that in any case the effects due to the variation of polarization and represented in Figures 3.14 and 3.15 are appreciated on the time scale of ms and can be compensated with the techniques already described.

Moreover we can see the sixteen representations as a natural coding of the electromagnetic field. Indeed every value of the quadruple (a, b, c, d) corresponds to different values of the triples (S_1,S_2,S_3), (T_1,T_2,T_3), (R_1,R_2,R_3), etc.

A transmission system able to switch pseudo randomly over different representation will make information unrecognizable by an attacker until he is able to reconstruct the pseudo-random sequence used by the transmitter and receiver of the transmission system. Figure 3.16 shows the effect of pseudo-random representation switching between some representations: one can understand how the continuous hopping from one representation to another, added to the "noises" accumulated during optical propagation, makes the signal unrecognizable, above all due to the impossibility of tracking the polarization variations due to the birefringence [26].

Such a system, that we could name a "representation-hopping" transmission systems, present the advantage respect to other coding schemes that all the representations are made from the same ten fundamental elements and thus the electronic implementation of transmitter and receiver is simpler than a generic coding. In fact the receiver described in [6] and better in [7] already extracts the values of a, b, c, d and carry out the products and the algebraic sums necessary to electronically implement (3.33), (3.34), (3.35), (3.36) is quite simple. More challenging – but also more efficient – could be modifying the receiver already described in [7] trying to extract directly the

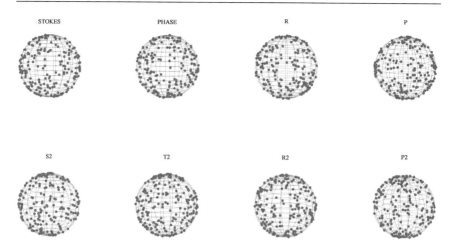

Figure 3.16 Effect of "representation hopping" on the received electromagnetic signal, observed in different representations.

ten fundamentals element without going through the calculation of the quadruple (a, b, c, d) and then the corresponding values for the sixteen representation.

Finally, we can state that the 4-Quadrature Signaling representation is the only one complete, or rather the only one able to exploit the whole information transfer capacity of the electromagnetic field. Other representations are projections in transformed spaces of lesser size, convey less information and can be combined to achieve a similar capacity to that of 4-Quadrature.

Moreover, the partial orthogonality of the identified spaces can be exploited to obtain new transmitting systems, using up to three spaces or alternatively taking into account different representations.

Finally, it is important to note that the 4 representations described – and all the 16 mathematically possible ones – constitute a natural encoding of the signal and accompany any electromagnetic signal.

REFERENCES

[1] C. E. Shannon, "Communication in the presence of noise," *Proceedings of the IRE*, vol. 37, pp.10–21, 1949.
[2] D. Slepian, "Permutation modulation," *Proceedings of the IEEE*, vol. 53, pp.228–236, 1965.
[3] R. Ottoson, "Group codes for phase- and amplitude-modulated signal on a Gaussian channel," *IEEE Transactions on Information Theory*, vol. 17, pp.315–321, 1971.
[4] G.R. Welti, J.S. Lee, "Digital transmission with Coherent four dimensional modulation," *IEEE Transactions on Information Theory*, vol. 20, 4, pp.497–502, 1974.

[5] J.M. Wozencraft, I.M. Jacobs, *Principles of Communication Engineering*, John Wiley & Sons, New York, pp.254–257, 1965.

[6] S. Betti, F. Curti, G. De Marchis, E. Iannone, "A Novel Multilevel Coherent Optical System: 4-Quadrature Signaling," *IEEE Journal of Lightwave Technology*, vol.9, n.4, pp.514–523, 1991.

[7] S. Betti, G. De Marchis, E. Iannone, "Multifrequency modulation for high-sensitivity coherent optical systems," *IEEE Journal of Lightwave Technology*, vol.11, n.11, pp.1839–1844, 1993.

[8] S. Betti, F. Curti, G. De Marchis, E. Iannone, "A Multilevel Coherent Optical System," Patent WO/1991/018455, November 1991. https://patentscope.wipo.int/search/en/detail.jsf?docId= WO1991018455&tab=PCTBIBLIO

[9] E. Agrell, M. Karlsson, "Power-efficient modulation formats in coherent transmission systems," *IEEE Journal of Lightwave Technology*, vol.27, n.22, pp. 5115–5126, 2009.

[10] D. Sasha, T. Birdsall, "Quadrature-quadrature phase-shift keying," *IEEE Transactions on Communications*, vol.37, n.5, pp.437–448, 1989.

[11] L. Zetterberg, H. Brändström, "Codes for combined phase and amplitude modulated signal in a four-dimensional space," *IEEE Transactions on Communications*, vol.COM-25, n.9, pp. 943–950, 1977.

[12] G. Taricco, E. Biglieri, V. Castellani, "Applicability of four-dimensional modulations to digital satellites: a simulation study," *Proceedings of GLOBECOM '93. IEEE Global Telecommunications Conference*, vol.4, pp.28–34, 1993.

[13] S. Betti, F. Curti, G. De Marchis, E. Iannone, "Exploiting fiber optics transmission capacity: 4-quadrature multilevel signalling," *Electronics Letters*, vol.26, n.14, pp.992–993, 1990.

[14] S. Betti, G. De Marchis, E. Iannone, P. Lazzaro, "Homodyne optical coherent systems based on polarization modulations," *IEEE Journal of Lightwave Technology*, vol.9, n.10, pp.1314–1320, 1991.

[15] R. Cusani, E. Iannone, A. Salonico, M. Todaro, "An efficient multilevel coherent optical system: M-4Q-QAM," *IEEE Journal of Lightwave Technology*, vol.10, n.6, pp.777–786, 1992.

[16] M. Karlsson, E. Agrell, "Spectrally efficient four-dimensional modulation," *OFC/NFOEC*, Los Angeles, CA, pp. 1–3, 2012.

[17] M. Karlsson, "Four-dimensional rotations in coherent optical communications," *IEEE Journal of Lightwave Technology*, vol.32, n.6, pp. 1246–1257, 2014.

[18] R. Jones, "A new calculus for the treatment of optical systems I. Description and discussion of the calculus," *Journal of Optical Society of America*, vol.31, n.7, pp.488–493, 1941.

[19] L. Arend, J. Krause, M. Marso, R. Sperber. "Four-dimensional signalling schemes – Application to satellite communications," ArXiv, November 2015.

[20] P. Perrone, S. Betti, G.G. Rutigliano, "A novel coherent multilevel combined phase and polarization shift keying modulation in twisted fibers," *Fiber and Integrated Optics*, pp.1–15, 2017.

[21] P. Perrone, S. Betti, G.G. Rutigliano, "Multidimensional modulation in optical fibers," *Research Journal of Optics and Photonics*, vol.2, n.1, pp. 1–8, 2018.

[22] M. Born, E. Wolf, *Electromagnetic theory of propagation, interference and diffraction of light. Principles of Optics*, Pergamon Press, Oxford, New York, 1959.

[23] M. R. Dennis, "Polarization singularities in paraxial vector fields: morphology and statistics," *Optics Communications*, vol.213, pp.201–221, 2002.

[24] G. G. Rutigliano, S. Betti, P. Perrone, "Representations of optical fiber communications in four and three dimensional spaces," *20th Italian National Conference on Photonic Technologies (Fotonica 2018)*, Lecce, Italy, pp.1–4, May 2018.

[25] G. G. Rutigliano, S. Betti, P. Perrone, "Representations of optical fibre communications in three- and four-dimensional spaces," *IET Communications*, vol.13, n.20, pp.3558–3564, 2019.

[26] G. G. Rutigliano, S. Betti, P. Perrone, "Multidimensional Secure Multilevel Polarization Shift Keying," *Fiber and Integrated Optics*, vol.35, n.5–6, pp.199–211, 2016.

Conclusions

Currently, optical and optoelectronic technologies make possible the realization of high-performance optical fiber communication systems and networks with the adoption of WDM configurations and both linear and nonlinear optical amplifications.

The last step for increasing the network throughput is represented by the implementation of multilevel modulation formats in coherent optical communication systems, which enable to increase the bit rate/channel, typically in WDM optical communication systems, over 100 Gbit/s/channel toward 400 Gbit/s/channel and beyond.

Within this prospect, novel conception multidimensional modulation formats together with non-conventional optical coherent systems can offer significant opportunities and advantages.

In particular, the 4-quadrature optical coherent communication system, proposed in 1990, based on a multilevel modulation in a four-dimensional (4-quadratures of the electromagnetic field in the optical fiber) physical space, can be assumed as "reference" since it has opened a novel perspective for multidimensional signaling in optical fiber in which information is represented in a four-dimensional space that is the maximum dimensionality of the physical space used for electromagnetic communications, for a given power value that is within a spherical symmetry.

This "reference" system has been then deeply analyzed in terms of power efficiency and its properties have been derived from the physical point of view starting from the fundamental quantum mechanical boson commutation relations and relating the system performance to the possible physical and non-physical degrees of freedom in four-dimensional space. As demonstrated by Karlsson, of the possible six degrees of freedom of a topological four-dimensional space, four are related to physical magnitudes while two have only mathematical meaning.

As general result, from the 4-Quadrature Signaling, all the binary and multilevel, non-coded, constant power modulation formats in two- and three-dimensional spaces can be derived by space rotations and compressions.

Therefore, the 4-Quadrature Signaling System can be either directly exploited for information transmission or used to derive modulation formats in three- or two-dimensional spaces.

For example, as regards the physical magnitudes, the three-dimensional spaces of the polarization (Poincaré Sphere) and the phase (Dennis Sphere) can be derived and also "hybrid" or non-physical spaces can be obtained.

This approach can be obviously used not only for information transmission but also for designing methods and/or algorithm for "signal protection" at physical level.

In order to pursue this goal, a fundamental question must be first of all clarified, that is, for a given constellation in four-dimensional space, how is it possible to derive the corresponding constellations in the sub-dimensional physical, hybrid and non-physical spaces?

In conclusion, the particular structure of the "reference" 4-Quadrature Signaling System opens new perspectives not only for information transmission but also for the "signal" protection at physical level, by unconventional "coding" techniques based on a different use of the constellation points and the effect due to the transition from the four-dimensional space to the sub-spaces. As a consequence, this operation could modify the information content for a given constellation so as to have an "entropic" effect, but, at the same time, it can provide a sort of "natural coding" without band expense.

Index